Studies in Computational Intelligence

Volume 586

Series editor

Janusz Kacprzyk, Polish Academy of Sciences, Warsaw, Poland
e-mail: kacprzyk@ibspan.waw.pl

About this Series

The series "Studies in Computational Intelligence" (SCI) publishes new developments and advances in the various areas of computational intelligence—quickly and with a high quality. The intent is to cover the theory, applications, and design methods of computational intelligence, as embedded in the fields of engineering, computer science, physics and life sciences, as well as the methodologies behind them. The series contains monographs, lecture notes and edited volumes in computational intelligence spanning the areas of neural networks, connectionist systems, genetic algorithms, evolutionary computation, artificial intelligence, cellular automata, self-organizing systems, soft computing, fuzzy systems, and hybrid intelligent systems. Of particular value to both the contributors and the readership are the short publication timeframe and the worldwide distribution, which enable both wide and rapid dissemination of research output.

More information about this series at http://www.springer.com/series/7092

Mincho Hadjiski · Nikola Kasabov
Dimitar Filev · Vladimir Jotsov
Editors

Novel Applications
of Intelligent Systems

 Springer

Editors
Mincho Hadjiski
Institute of Information and Communication
  Technologies
Bulgarian Academy of Sciences
Sofia
Bulgaria

Nikola Kasabov
Knowledge Engineering and Discovery
  Research Institute (KEDRI)
Auckland University of Technology
Auckland
New Zealand

Dimitar Filev
Research and Advanced Engineering
Ford Motor Company
Dearborn, MI
USA

Vladimir Jotsov
University of Library Studies and
  Information Technologies
Sofia
Bulgaria

Studies in Computational Intelligence
ISBN 978-3-319-79194-4   ISBN 978-3-319-14194-7   (eBook)
DOI 10.1007/978-3-319-14194-7

Printed on acid-free paper

This Springer imprint is published by SpringerNature
The registered company is Springer International Publishing AG Switzerland

Preface

In this book "Novel Applications of Intelligent Systems," selected scientific and scientific-application investigations on intelligent systems are presented. The problems vary from industrial to web and problem-independent applications. All this is united under the slogan: "Intelligent systems conquer the world." It is difficult today to find innovation projects without any analytical research, invention, retrieval and processing of knowledge and logical applications in technology. The crisis that burst forth in 2008 and the following recession decelerated slightly the pace of intelligent applications, but today some countries are preparing for a technological jump on the base of the innovation technologies stored, while others will quickly lag.

That is why this book is recommended to a wide circle of readers and it is particularly recommended to the young generation of IT experts who will build the next generations of intelligent systems.

In the chapter "Modern Approaches for Grain Quality Analysis and Assessment" authors M. Mladenov, M. Deyanov, and S. Penchev present methods and tools for assessment of the main quality features of grain samples, which are based on color image and spectral analysis. The results obtained by two proposed data fusion approaches are compared.

In "Intelligent Technical Fault Condition Diagnostics of Mill Fan" written by M. Hadjiski and L. Doukovska, sase-based reasoning (CBR) approach is used where predictive modeling and the standard statistical applications do not provide satisfactory results. Positive effects of the various machine self-learning methods application are also investigated

In "Abstraction of State-Action Space Utilizing Properties of the Body and Environment" by K. Ito, S. Kuroe and T. Kobayashi, innovative results are presented for industrial applications of autonomous robotized systems. Reinforcement learning is used in the method being proposed for control of the different types of robots developed. As a result, the lack of reproducibility problem and other important problems have been solved.

K. Shiev, S. Ahmed, N. Shakev, and A. Topalov are the authors of the next, chapter "Trajectory Control of Manipulators Using an Adaptive Parametric Type-2 Fuzzy CMAC Friction and Disturbance Compensator." They elaborated a novel type-2 fuzzy cerebellar model articulation controller neural network aiming at better trajectory tracking robot control. An overview of the manipulator dynamics and CT control is given. The control approach guarantees the stability of closed-loop system.

In "On Heading Change Measurement: Improvements for Any-Angle Path-Planning" by P. Muñoz and M. D. R-Moreno, an original research is presented for finding the most efficient and safe path between locations using mobile robots and other applications.

In the next chapter, the authors G. Ulutagay and E. Nasibov present their research "C × K-Nearest Neighbor Classification with Ordered Weighted Averaging Distance." Ordered weighted averaging (OWA) distance is used in a modification of K-nearest neighbors method. By just adjusting the OWA weights the authors show different clustering strategies.

P. Sadeghi-Tehran and P. Angelov wrote the chapter "ARTOD: Autonomous Real Time Objects Detection by a Moving Camera Using Recursive Density Estimation," where a new approach to autonomously detect moving objects in a video captured by a moving camera is proposed. Two surveillance videos were tested which can be categorized as ground-based and aerial videos.

Improved genetic algorithm is used for optimization of the search space by N. El-Zarif and M. Awad in order to solve the problem of downlink resource allocation in a special class of networks. The research in the chapter "Improved Genetic Algorithm for Downlink Carrier Allocation in an OFDMA System" is dedicated to this problem.

In the chapter "Structure-Oriented Techniques for XML Document Partitioning ," G. Costa and R. Ortale consider data mining approaches focusing on data clusterization from XML texts. Two new approaches are proposed in this connection whose effectiveness is demonstrated by comparative analysis with the other best representatives in the area. An experimental evaluation of the devised techniques is presented.

Chapter "Security Applications Using Puzzle and Other Intelligent Methods" by V. Jotsov and V. Sgurev is dedicated to applications of special types of data analytics methods called puzzle methods. The name of this group of methods is chosen due to the fact that some of the new types of constraints introduced in this direction may be explained by analogy with human thinking when solving su-doku, puzzles, etc. It is shown that application of this group of methods for the purpose of security increases considerably the indicators of actions' unexpectedness and the applications' actions and protection security as a whole.

In the chapter "Semiautomatic Telecontrol by Multi-link Manipulators Using Mobile Telecameras" written by V. Filaretov and A. Katsurin, a system for semiautomatic control of submarine robots or manipulators is presented. This investigation resolves the problem of telecontrol by multi-link manipulator when the telecamera orientation changes. Original solutions are also proposed for the problem of developing new methods and algorithms of semiautomatic telecontrol by multi-link manipulators with changing of orientation the optical axis of television camera, which is located in the zone of the realization of working operations. This paper resolves the problem of developing two methods of semiautomatic position and combined telecontrol by multi-link manipulators using the setting devices whose kinematic schemes are differed from the kinematic schemes of manipulators.

Map-building and localization are some of the fundamental topics in mobile robots research. S. Rady in "Vision-Based Hybrid Map-Building and Robot Localization in Unstructured and Moderately Dynamic Environments" focuses on developing efficient environment representation and localization for mobile robots. The solution-approach is suitable for unstructured and moderately dynamic environments. The approach proposed is capable of localizing a moadaptive neurobile robot at both topological and metric levels. A hybrid map construction is proposed, based on maintaining distinctive features to provide recognition accuracy with less computational overhead and the map size is simultaneously decreased.

"Innovative Fuzzy-Neural Model Predictive Control Synthesis for Pusher Reheating Furnace" by G. Stojanovski, M. Stankovski, I. Rudas, and J. Jing is dedicated to the fuzzy-neural variant of the Sugeno fuzzy model. It is used as an adaptive neuro-fuzzy implementation and employed as a predictor in a predictive controller. In order to build the predictive controller the adaptation of the fuzzy model using dynamic process information is carried out. Simulation results for RZS Furnace at Skopje Steelworks are also presented. This control system employs a fuzzy-neural model to implement the predicting function and a gradient-optimization algorithm to synthesize the controlling sequence and close the control loop.

In "Exactus Expert—Search and Analytical Engine for Research and Development Support" written by G. Osipov, I. Smirnov, I. Tikhomirov, I. Sochenkov, and A. Shelmanov the system Exactus Expert is presented—a search and analytical engine. This is a kind of analytical tool that can process large amounts of unstructured and semistructured data, which is basically represented by texts in different natural languages. The search and analytic engine "Exactus Expert" is demanded by experts to support the decision-making process on research topics funding by giving aggregated information about different sides of scientific activity.

In the last chapter of the book "Acoustic and Device Feature Fusion for Load Recognition" the authors A. Zoha, A. Gluhak, M. Nati, M. Ali Imran, and S. Rajasegarar discuss the initial investigation of a multi-layer decision framework for smart energy sensing. Their objective is to improve the device recognition accuracy of low-power consumer appliances by combining steady-state load features with audio features derived from the device usage. They investigate the use of time-domain and FFT-based audio feature sets for recognizing acoustic activity within an office environment. SVM was found out to be the best classification model for both audio and device recognition tasks.

September 2014

Mincho Hadjiski
Nikola Kasabov
Dimitar Filev
Vladimir Jotsov

Contents

Modern Approaches for Grain Quality Analysis and Assessment 1
M. Mladenov, M. Deyanov and S. Penchev

Intelligent Technical Fault Condition Diagnostics of Mill Fan 23
Mincho Hadjiski and Lyubka Doukovska

**Abstraction of State-Action Space Utilizing Properties
of the Body and Environment** 41
Kazuyuki Ito, So Kuroe and Toshiharu Kobayashi

**Trajectory Control of Manipulators Using an Adaptive Parametric
Type-2 Fuzzy CMAC Friction and Disturbance Compensator** 63
Kostadin Shiev, Sevil Ahmed, Nikola Shakev and Andon V. Topalov

**On Heading Change Measurement: Improvements
for Any-Angle Path-Planning** 83
Pablo Muñoz and María D. R-Moreno

**C × K-Nearest Neighbor Classification with Ordered Weighted
Averaging Distance** 105
Gozde Ulutagay and Efendi Nasibov

**ARTOD: Autonomous Real-Time Objects Detection by a Moving
Camera Using Recursive Density Estimation** 123
Pouria Sadeghi-Tehran and Plamen Angelov

**Improved Genetic Algorithm for Downlink Carrier Allocation
in an OFDMA System** 139
Nader El-Zarif and Mariette Awad

Structure-Oriented Techniques for XML Document Partitioning 167
Gianni Costa and Riccardo Ortale

Security Applications Using Puzzle and Other Intelligent Methods 183
Vladimir Jotsov and Vassil Sgurev

**Semiautomatic Telecontrol by Multi-link Manipulators
Using Mobile Telecameras** 203
Vladimir Filaretov and Alexey Katsurin

**Vision-Based Hybrid Map-Building and Robot Localization
in Unstructured and Moderately Dynamic Environments** 231
Sherine Rady

**Innovative Fuzzy-Neural Model Predictive Control Synthesis
for Pusher Reheating Furnace** 251
Goran S. Stojanovski, Mile J. Stankovski, Imre J. Rudas and Juanwei Jing

**Exactus Expert—Search and Analytical Engine for Research
and Development Support** 269
Gennady Osipov, Ivan Smirnov, Ilya Tikhomirov, Ilya Sochenkov
and Artem Shelmanov

Acoustic and Device Feature Fusion for Load Recognition 287
Ahmed Zoha, Alexander Gluhak, Michele Nati, Muhammad Ali Imran
and Sutharshan Rajasegarar

Modern Approaches for Grain Quality Analysis and Assessment

M. Mladenov, M. Deyanov and S. Penchev

Abstract The paper presents the approaches, methods and tools for assessment of main quality features of grain samples which are based on color image and spectra analyses. Visible features like grain color, shape, and dimensions are extracted from the object images. Information about object color and surface texture is obtained from the object spectral characteristics. The categorization of the grain sample elements in three quality groups is accomplished using two data fusion approaches. The first approach is based on the fusion of the results about object color and shape characteristics obtained using image analysis only. The second approach fuses the shape data obtained by image analysis and the color and surface texture data obtained by spectra analysis. The results obtained by the two data fusion approaches are compared.

1 Introduction

It will not be forcedly to say, that one of the main factor of human life quality is the food quality and safety. The food provides the energy, needed for the human body for movement, physical and intellectual activity. It is a source of proteins, fats, carbohydrates, vitamins and minerals, due to them the cells and tissues are renovated. As a result of the feeding the human organism produces hormones, enzymes and other regulators of the metabolic processes.

The assessment of food product quality and safety is an important part of food production chain. The grain is very important part of the food for the humans, as well as for the animals. The higher food quality requirements require development

M. Mladenov (✉) · M. Deyanov · S. Penchev
Department of Automatics and Mechatronics, Faculty of Electrotechnics,
Electronics and Automatics, University of Rousse, 8 Studentska Str., 7017 Rousse, Bulgaria
e-mail: mladenov@uni-ruse.bg
URL: http://www.uni-ruse.bg

© Springer International Publishing Switzerland 2016
M. Hadjiski et al. (eds.), *Novel Applications of Intelligent Systems*,
Studies in Computational Intelligence 586, DOI 10.1007/978-3-319-14194-7_1

of new, objective, intelligent technologies, methods and tools for assessment of main food quality and safety features.

The grain and the grain foods are the basic components of the human food. The grain foods assure the half of the daily energy ration of the people in the developed countries and 80 % in the developing countries. The grains of the wheat, maize, rice, barley, oats and millet content about 60–80 % carbohydrates, 7–14 % proteins and 2–8 % fats.

The task for rapid, objective, automated, express and non destructive assessment of the grain quality is a complex and multilevel task, related to an analysis of the common appearance and the visible features of the grain sample elements, as well as with the grain contents, smell, flavour, moisture content, infections, non grain impurities, etc.

This investigation is focused on the assessment of main quality features of corn grain. It is proposed and investigated a complex approach, methods and tools for feature extraction and data dimensionality reduction, analysis and identification of the grain sample elements. They are based on the analysis of color images and spectral characteristics of the grain sample elements, as well as the fusion of the results of two kind of analyses. This approach is realized in the frames of INTECHN project "Development of Intelligent Technologies for Assessment of Quality and Safety of Food Agricultural Products" founded by Bulgarian National Science Fund.

According to the Bulgarian national standards the main quality features of grain are: inherent for the variety appearance, shape, color, smell, taste, moisture and impurities. The main quality features of a grain corn sample are presented in Table 1.

Besides the features presented in Table 1, which are mainly related to the visible characteristics of the grain sample elements, characteristics related to the grain composition, dry matter content, moisture content, starch, protein, glutenin, vitamins, toxins and mineral content are used in the assessment of grain quality.

Table 1 Corn grain quality groups

Grain quality groups	Grain quality features
First group—standard kernel	Whole grains and broken grains bigger than the half of the whole grain, with appearance, shape and color inherent for the variety
Second group—grain impurities	Broken grains smaller than the half of the whole grain, heat-damaged grains, small grains, shriveled grains, green grains, sprouted grains, infected (with Fusarium) grains, smutty grains
Third group—non grain impurities	Corn-cob particles, leaf and stem fractions, pebbles, soil and sand, as well as harmful elements

2   Modern Approaches for Grain Quality Assessment

It is obvious that all of the characteristics mention above can not be evaluated on the basis of one and only approach, based on the information extracted from one separated sensor source. Because of a big part of these characteristics are evaluated by an expert on the basis of visual assessment only, it is appropriate a computer vision system to be used for such characteristics extraction and evaluation. Another part of the grain features, which are related to the grain composition, grain content, grain infections etc., on principle can not be evaluated through image analysis. Effective results in the assessment of these grain features are obtained using grain spectral characteristics analysis. Another part of the grain features, like moisture content, specific weight etc., are evaluated by precise standard methods for physical chemistry analyses.

2.1   Grain Quality Features Assessment Using Color Image Analysis

Most of the grain quality characteristics specified in Table 1 (except moisture content, smell and flavour) are related to the colour characteristics, shape and dimensions of the grain sample elements. The main trend in the last few years has been to use Computer Vision Systems (CVS) in order to evaluate such characteristics. A review of the progress of computer vision in the agricultural and food industry is given in [2]. To obtain a complex assessment of the grain quality using data about colour characteristics, shape and dimensions of the grain sample elements, is a complicated and multilevel task [16]. This is because the colour characteristics, the shape and the dimensions of the elements in a sample vary within a wide range.

Many results are published, in which colour image analysis is used to assess some particular quality features. In a digital image analysis algorithm [13] is developed to facilitate classification of individual kernels of cereal grains using textural features of individual grains. Colour characteristics analysis is used also to assess variety [13, 14], infections [20, 23], germination [18], weed identification [1], etc.

Morphological features, related to the grain shape and geometrical parameters are used for assessment of the grain variety. A set of eight morphological features namely, area, perimeter, length of major axis, length of minor axis, elongation, roundness, Feret diameter and compactness are used [24] to recognize five different kinds of cereal grains. A broader investigation, with a total of 230 features (51 morphological, 123 colour, and 56 textural) used for classification of barley, Canada Western Amber Durum wheat, Canada Western Red Spring wheat, oats, and rye is presented in [25]. Assessment of the grain sample purity is performed by profile analysis of corn kernels using one-dimensional digital signals based on its binary images [9], by modeling the shape using a set of morphological features [26] and by shape curvature analysis [16]. Computer vision methods are also used for determination kernel mechanical damage and mold damage [22], broken kernels in threshing process [32], etc.

2.2 Grain Quality Features Assessment Using Spectra Analysis

Some preliminary investigations [20] showed, that we can't get sufficiently precise assessment of some of grain sample elements like smutty grains, infected grains and non-grain impurities using image analysis only. This is conditioned by the fact that a change of the color characteristics, as well as of the surface texture is appeared for these grain sample elements. It is difficult to detect the change of surface texture using CVS. That's why we expect a more accurate assessment of these sample elements to be obtained using spectra analysis. Unfortunately information about object shape and dimensions can not be extracted from spectra.

Substantial investigation, in which Visible (VIS) and near infrared (NIR) spectra analyses are used for evaluation of quality features of different food products are published in [3, 16, 21, 33]. VIS and NIR spectra analyses are applied in the assessment of different grain quality features too [4–7, 16]. They are mainly used in tasks, related to the determination of qualitative and quantitative features like grain composition, dry matter content, moisture content, starch, protein, glutenin, vitamins, toxins, mineral content, etc. Different calibration models are developed for predicting grain composition and content [4, 15, 27–29, 35]

Modified partial least squares models on NIR spectra (850–1048.2 nm) are developed for each constituent or physical property [15]. The best models are obtained for protein, moisture, wet gluten, and dry gluten with $r^2 = 0.99, 0.99, 0.95$, and 0.96, respectively.

The spectra analysis is used for detection of different grain infections. Determination and prediction of the content of ergosterols and different kinds of mycotoxins like aflatoxin, fumonisin and others are very important tasks because mycotoxins are toxic for animals and humans. Reflectance and transmittance VIS and NIR spectroscopy are used to detect fumonisin in single corn kernels infected with *Fusarium verticillioides* [4]. A method for determination of *Fusarium graminearum* infection is proposed in [8]. The classification accuracy is up to 100 % for individual samples. Transmittance spectra (500–950 nm) and reflectance spectra (550–1700 nm) are used as tools for aflatoxin determination in single whole corn kernels [29]. The authors used discriminant analysis and partial least squares regression for spectral data processing. The best results are obtained using two feature discriminant analyses of the transmittance data. A NIRS method for estimation of sound kernels and Fusarium-damaged kernels proportions in grain and for estimation of deoxyinivalenol levels is proposed in [30]. The method is classified sound and Fusarium damaged kernels with an accuracy of 98.8 and 99.9 %, respectively. A neural network based method is developed for deoxynivalenol levels determination in barley using NIRS from 400 to 2400 nm [31]. Fourier transform NIRS is applied for rapid and non-invasive analysis of deoxynivalenol in durum and common wheat [6]. A qualitative model for discrimination of blank and naturally contaminated wheat samples is developed. Classification accuracy of the model is 69 % of the 65 validation samples.

2.3 Grain Quality Features Assessment Using Hyperspectral Analysis

A comparatively new approach for grain quality assessment is based on the Hyperspectral Imaging System (HIS). A HIS forms data about object spectra in a set of regions (pixels) of the object area. Every pixel contents information about spectral signature of reflection for a large number of narrow spectral bands over a continuous spectral range (usually in VIS and NIR spectrum). Spectral data is normally presented as a hyperspectral cube. In a sense a HIS could be considered as a variant of a color image analysis, where the object image is divided into pixels, every pixel is analyzed using multiband spectral analysis instead of three band analysis.

The hyperspectral analysis is applied in the assessment of different grain features. For example, Mahesh et al. [12] used HIS for developing of class models of different wheat varieties in Western Canada. The grain samples are scanned in NIR spectrum (960–1700 nm) at an interval of 10 nm. 75 different values of the intensity of the reflection are extracted from hyperspectral images and they are used for class model development. These models assure about 90 % classification accuracy.

NIR spectroscopy is applied for assessment of grain moisture level too [10, 11]. The authors presented a new method using NIR hyperspectral imaging system (960–1700 nm) to identify five western Canadian wheat classes at different moisture levels. They are found that the linear discriminant analysis (LDA) and quadratic discriminant analysis (QDA) could classify moisture contents with classification accuracies of 89–91 and 91–99 %, respectively, independent of wheat classes. Once wheat classes are identified, classification accuracies of 90–100 and 72–99 % are observed using LDA and QDA, respectively, when identifying specific moisture levels.

The HIS (350–2500 nm) is used for assessment of protein content in wheat grains [34] too.

3 Assessment of Basic Grain Quality Features Using CVS

3.1 Grain Groups and Subgroups Using Color Image Analysis

Such characteristics of grain sample elements, which are on principle evaluated by expert on the basis of visual estimation, are assessed using CVS in the frame of this investigation. They are related with the appearance, the color, the shape and the dimensions of the grain sample elements. Groups (classes) and subgroups (sub-classes) in which have to be distributed corn grain sample elements are presented in Table 2.

Because of the color and shape features are described in a different manner, it is advisability these characteristics assessment has to be made separately. After that the results from two assessments have to be fused to obtain the final assessment about object categorization to one of the normative classes. In terms of the classification procedure color and shape groups are divided in several subgroups. Color object characteristics are separated in 8 basic classes corresponding to the typical for the different sample elements color zones and 1 additional class that corresponds to the non-grain impurities (it is impossible to define a compact class for non-grain impurities). According to the object shape the objects are divided in 3 basic classes corresponding to the whole grains, broken grains bigger than the half of whole grain and broken grains smaller than the half of whole grain, and 1 additional class, that corresponds to the non-grain impurities. Each of the three basic shape classes is divided in 6 shape subclasses.

Table 2 Corn grain sample classes and subclasses

Normative (quality) classes	Color classes	Shape classes
1cst—standard kernel (whole grains and broken grains bigger than the half of the whole grain,) with appearance, shape and color inherent for the variety	1cc—grains with color inherent for the variety, back side	1csh—whole grains with inherent for the variety shape
	2cc—grains with color inherent for the variety, germ side	
	3cc—heat-damaged grains	2csh—broken grains bigger than the half of the whole grain
2cst—grain impurities: broken grains smaller than the half of the whole grain, heat-damaged grains, small grains, shriveled grains, green grains, sprouted grains, infected (with Fusarium) grains, smutty grains	4cc—green grains	3csh—broken grains smaller than the half of the whole grain and small and shriveled grains
	5cc—mouldy grains	
	6cc—smutty grains	
3cst—non grain impurities: corn-cob particles, leaf and stem fractions, pebbles, soil and sand, as well as harmful elements	7cc—infected (with Fusarium) grains	
	8cc—sprouted cgrains	4csh—non-grain impurities
	9cc—non-grain impurities	

3.2   Features Extraction from Images

RGB, HSV, XYZ, NTSC and YCbCr color models are used for separating object area from background and for different color zones extraction in the frame of object area (inherent for the variety color, heat-damaged grains, green grains, smutty grains, infectious (with Fuzarium) grains, bunt, non-grain impurities). Furthermore four color texture models [19] are development for this purpose. It is expected they will better underline zone color ratios in the input RGB image. The texture models can be presented by the following equations:

First texture model. It is constructed on the basis of RGB model. Its components are the normalized differences between the R, G and B components:

$$d_1 = \frac{R - G}{R + G + B}; d_2 = \frac{B - R}{R + G + B}; d_3 = \frac{G - B}{R + G + B}; d_1 + d_2 + d_3 = 0 \quad (1)$$

Second texture model. This model includes non-linear transformation onto R, G and B components as follows:

$$O_1 = 1\frac{R}{G}; O_2 = \frac{B}{R}; O_3 = \frac{G}{B}; O_1 O_2 O_3 = 1 \quad (2)$$

Third texture model. It is similar to the second model. The pixel intensity is added as fourth coordinate:

$$O_1 = \frac{R}{G}; O_2 = \frac{B}{R}; O_3 = \frac{G}{B}; I = R + G + B; O_1 O_2 O_3 = 1 \quad (3)$$

Fourth texture model. This model converts input RGB space into one-dimensional texture feature T_k:

$$T_k = \frac{3(R + mG + nB)}{(R + G + B)(1 + m + n)}; m = \frac{G}{R}; n = \frac{B}{R} \quad (4)$$

For separation of the object region from the background, the best results are obtained using the second texture model. When extracting different color zones within the object region, the best performance have the HSV, RGB color models, as well as the first and second texture model. Therefore, the operator, who performs the INTECHN platform training, has the possibility to choose the appropriate color or texture model based on the results obtained during the training procedure.

To represent the shape of the investigated objects 10—dimensional vector descriptions are used [16, 17]. The following approach is used to obtain shape description of the investigated objects. First, the binary image of the object zone is created. After that the object's peripheral contour is extracted, the bisection line of

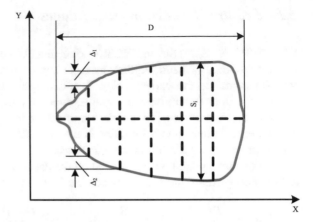

Fig. 1 Object shape description: D—length of the bisection line; $h_i = s_i/D$—length of a cross-section; $\Delta_i = \Delta_1 - \Delta_2$—difference between neighbor cross-sections

the contour is detected and an odd number of cross-sections perpendicular to the bisection line are built (Fig. 1).

The cross-sections relative length $h_i = s_i/D$, as well as the size and the sign of the difference between two neighbor sections $\Delta_i = \Delta_1 - \Delta_2$ are calculated. Finally the object shape description is presented in the following form:

$$X_{sh} = (h_1, h_2, \ldots, h_n, \Delta_1, \Delta_2, \ldots, \Delta_n) \tag{5}$$

It is typical for maize kernels that its contour line has a huge asymmetry at the part where the germ exists and also at the opposite part. It is easy to locate the germ in the whole grains and to build contour descriptions and models with proper orientation. For broken grains, depending on what part of the whole is remaining (with the germ or without it) the contour descriptions could be sufficiently different. That is why it is necessary for classes 2csh and 3csh (to be more precise —for its corresponding subclasses) to define two types of descriptions and models: for shapes where the germ exists in the remaining part of the grain and where it is not. The all shapes of the investigated objects are divided into 18 groups (subclasses).

4 Assessment of Basic Grain Quality Features Using Spectra Analysis

4.1 Grain Groups and Subgroups Using Spectra Analysis

Grain features related to color characteristics and surface texture are evaluated on the basis of spectra analyses in this investigation. This is conditioned by the fact, that for some grain sample elements, like Fusarium infected grains, shriveled grains, sprouted grains, smutty grains and non-grain impurities, the color image analyses doesn't obtained sufficiently precise data of features, related to the surface texture characteristics.

Groups and subgroups in which have to be distributed corn grain sample elements are presented in Table 2, section Color classes.

4.2 Features Extraction from Spectra and Data Dimensionality Reduction

Different methods like Principal Component Regression, Partial Least Squares Regression, Principal Component Analysis (PCA), Hierarchical Cluster Analysis and other methods are used for developing a model to predict a property of interest, as well as for feature extraction and large and complex spectra data reduction. Methods like K-Nearest Neighbors (KNN), Linear Discriminant Analysis (LDA), Quadratic Discriminant Analysis (QDA), Cluster Analysis (CA), Support Vector Machines (SVM), Neural Networks (NN), and Soft Independent Modeling of Class Analogy (SIMCA) are used for assessment of different grain features using data from grain spectra.

The classification of grain sample elements on the basis of their spectral characteristics are made in groups corresponding to the color classes presented in Table 1. The spectral characteristics of grain sample elements are obtained using QE65000 spectrophotometer. Each characteristic is a vector with about 1500 components. Principle Component Analysis (PCA) and combination of Wavelet descriptions and PCA are used for extracting typical features from object spectral characteristics and for spectral data dimensionality reduction. The following Wavelet coefficients are used: Wavelet1—detail coefficients and Wavelet2—approximation coefficients. The operator can select one of the following wavelet functions: Haar, Daubechies2, Coiflet2, Symlet2. The level of decomposition can vary from m = 1 to m = 4. The most informative wavelet coefficients are chosen using PCA method.

5 Grain Quality Assessment Fusing Data from Image and Spectra Analyses

Because of the fact that color and shape features are described in a different manner, the assessment of these characteristics is made separately. After that the results from the two assessments have to be fused in order to obtain the object's final categorization to one of the normative classes.

Different variants for fusing the results obtained from objects color characteristics and shape analyses were developed at different stages of the investigation [16]. The algorithms developed could be associated with hierarchical clustering

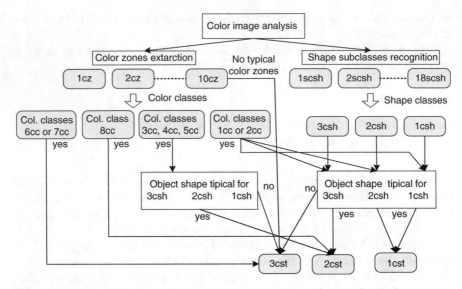

Fig. 2 Fusing the results from color characteristics and shape analyses. Variant 1

algorithms. Their typical feature is that different criteria for class merging are used at different levels of data fusion.

Variant 1. A comparatively simplified fusion scheme is used in the first algorithm. It is presented in Fig. 2. The input data (input classes) are separated in two groups—color characteristics data and objects shape data. The first group contains of 10 color zones inherent for the maize grains. The second group consists of 18 shape subclasses (1scsh, 2scsh,..., 18scsh). The first six of them correspond to different shape models of whole kernels, the next six—to models of broken grains bigger than the half of whole grain and the last six—to models of broken grains smaller than the half of whole grain.

The color class (1cc, 2cc,..., 8cc) is determined on the basis of preliminary defined combinations of color zones at the first stage of fusing the results from the color analysis. The shape subclasses are merged into one of the three main shape classes (1csh, 2csh and 3csh).

At the second stage of the analysis the fusion of color and shape classes is made in order to form the final decision of object classification in one of the three normative classes (1cst, 2cst and 3cst). The assessment whether the shape of the object is typical for one of the three main grain classes or not is used as a fusion criterion for color classes 1cc to 5cc. For 6cc, 7cc and 8cc classes the shape is not important at all.

Variant 2. The second algorithm (Fig. 3) uses color and combined topological models of typical color zones. The topological models represent the plane

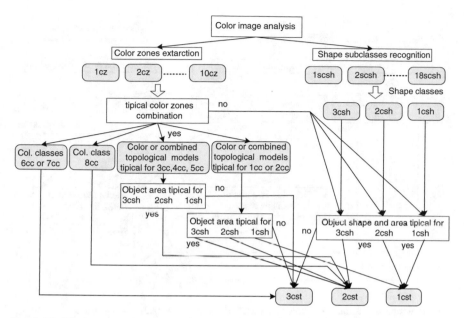

Fig. 3 Fusing the results from color characteristics and shape analyses. Variant 2

distribution of the color zones within the object area. A set of color topology models (when 3 or more typical for the kernels color zones are found) and combined topology models (when only 2 typical color zones are found) is preliminary defined. The combined topological models represent the plane distribution of some shape element (kernel prickle or rear area) and the color zones found.

The final objects categorization when such topology is found, is performed without taking into account the object's shape but only its area. The shape and area are the most important for the final categorization of the objects when only one typical color zone is found. When the object belongs to 6cc, 7cc or 8cc its shape is not important at all for taking the decision making.

Variant 3. The third data fusion algorithm (Fig. 4) is based on data about object color characteristics obtained through analysis of their spectral characteristics and data about object shapes extracted from object images. This variant of color and shape data fusion is conditioned by the fact that the recognition of color class of grain sample elements using spectral data is more precise.

The main criterion for final categorization is the object color class. The correspondence of object shape and/or area to the typical for the different shape classes shape and area is an additional criterion.

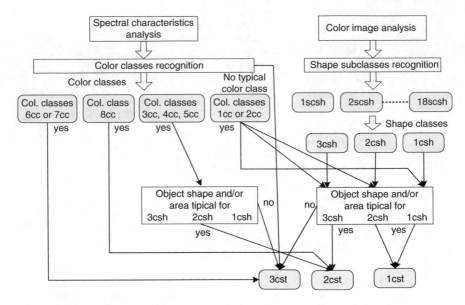

Fig. 4 Fusing the results from color characteristics and shape analyses. Variant 3

6   Classification of the Grain Sample Elements

Specific classification strategy, classifiers, and validation approach [16, 17] are used for the categorization of the grain sample elements, which are conditioned by the specificity of the classification tasks.

Classification approach. If the classes (related with the color, shape, PCA and Wavelet + PCA descriptions) are presented in the feature space, a part of them (1cc, 2cc,...8cc) will form comparatively compact class regions. The sets of descriptions extracted from the grain sample training sets are used for developing the models of these grain sample groups. Each class model is presented by the class centre (the average value of the class training data) and the class boundary surface. The boundary surface is determined through a threshold value of the covariance of the class training data. A correct model for the 9cc class (non-grain impurities) could not be created because of the fact that the characteristics of elements of this class could be sufficiently different in each subsequent grain sample.

As a correct model for the 9-th grain group could not be created, a part of the descriptions of such objects from the testing set could get into the boundaries of the other eight classes defined. A big part of them would get outside the class regions and could be located in a random place in the feature space. These descriptions could be considered as noisy vectors. It could be assumed that the comparatively compact class regions of the objects from the first eight groups are submerged in a noisy environment. Therefore, the task for categorization of the grain sample elements can be interpreted as a task for classification in classes, whose boundaries have definite shapes, dimensions and location in the feature space, and they are situated in a noisy environment [17].

Under this formulation, the use of popular strategies like LDA, CA, SVM, KNN and some other methods, which build boundaries between class regions, is obviously not a good choice. This is due to the fact that for the class 9cc, which correspond with the 9-th grain group, a correct model cannot be created.

Furthermore if there are too big deviations of the actual values of the object characteristics and intensive measurement noise, the class areas can be overlapped. Very often correct information about priory probabilities of the classes is missing which makes the classification problem more complex. If we use a classifier, which demands a priory probabilities to be known (for example Bayesian classifier), the training procedure has to be implemented using the a priory probabilities obtained from the number of elements in training sets. When we assess quality of an unknown sample, the ratio of the number of elements from different classes can be sufficiently different from this ratio in training sets. The classifier decision can be sufficiently different from the optimal decision under these circumstances. In this case the classification task is reduced to a task for overlapping class areas approximation when the classes are situated in noisy environment and correct information for class a priory probabilities is missing.

Classifiers. The task for grain class modeling is reduced to a task for approximation of the grain class regions. For this purpose, classifiers based on Radial Basis Elements (RBEs) are used. Such classifiers are chosen in terms of the simplicity of the classification procedure and the accuracy of the class region approximation. Furthermore, if we set an appropriate value of the RBE bias and a minimal threshold Δ of its output, it becomes clear what part of input vectors will be included within the class boundary and it is easy to change the dimensions of the particular class region.

The following classifiers [16] are used for class areas approximation.

Classifier 1 (CSRBE). Only one RBE is used for approximation of each class area (Fig. 5). The RBEs centers correspond to the class average values obtained from class training sets. The following notations are used: $f_{\omega i}$ is the output of ith RBE, which corresponds to the class ω_i; Δ is the threshold value which limits class area dimensions.

In the case when one standard RBE is used for each class area approximation the class boundaries are round shaped. To determine to what class the input vector belongs, the output $f_{\omega i}$ with maximum value is chosen. This value has to exceed the threshold value Δ of the threshold element (the last element of the classifier architecture).

Classifier 2 (CDRBE). Another solution for class area approximation gives the classifier architecture (CDRBE), presented in Fig. 6.

The first classifier layer consists of m transforming elements, which recalculate input vectors coordinates in local coordinate systems which axes coincide with the class axes of inertia.

The next layer consists of n x m RBEs, which are distributed in m sub layers (m is the number of classes). The number of RBEs in each sub layer is equal to n (n is

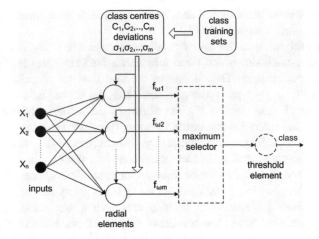

Fig. 5 Round shaped class areas approximation using standard RBEs (CSRBE)

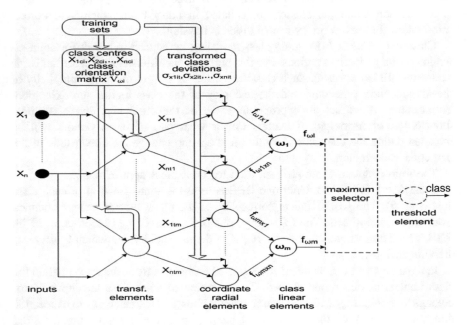

Fig. 6 Classifier with decomposing RBEs (CDRBE)

the input vectors dimensionality). Each RBE has one input, connected with one of the input vector coordinate. The RBEs centers coincide with average values of the projections of training sets vectors onto corresponding coordinates of class local coordinate system. The RBEs biases are set in conformity with standard deviations of the respective input vectors coordinates projections ($\sigma_{x1\omega i}$, $\sigma_{x2\omega i}$...$\sigma_{xn\omega i}$).

The third layer consists of m RBEs. The weights of all RBEs are equal (1, 1… 1). The RBEs outputs are the weighted distances of input vector to the centers of non-spherical classes.

The classifier architecture presented above gives a possibility to form classes, which dimensions along the directions of separate coordinate axes are different. Changing the RBEs biases and the threshold value Δ we can vary the class shape from sphere to shape close to parallelepiped.

Classifier 3 (**CRBEP**). It is well-known, that the optimal solution of the task concerning classification of vectors into overlapping classes can be obtained on the basis of the Bayes classification rule. Its application requires the conditional probabilities $P(X|\omega_i)$, as well as the prior probabilities $P(\omega_i)$ to be known. The condition for correct classification of an input vector X into the class ω_i can be given by the inequality:

$$P(X|\omega_i)P(\omega_i) > P(X|\omega_j)P(\omega_j), \quad j = 1\ldots m, \, j \neq i \qquad (6)$$

where m is the number of classes.

This variant of the Bayesian classifier creates boundaries between classes, but it can't approximate the class areas. Furthermore, if the ratio of the number of elements in training sets is different from the ratio of the number of elements in real classes (the same problem occurs in the discussed classification tasks), the Bayes classifier loses its optimality.

Let us suppose that the requirement for class areas approximation is valid and the ratio of numbers of elements in classes is changing during classification of unknown sample. Under this formulation the following approach is realized in the frames of the INTECHN platform. The class areas are approximated using standard (or decomposing) RBEs and the accumulated during classification number of vectors of each of the classes is interpreted as class potential $V_{\omega i}$. This potential is defined as:

$$V_{\omega_i} = N_i/N_{\max} \qquad (7)$$

where N_i is the number of vectors, which are currently classified in class ω_i; $N_{\max}$ is the number of vectors, which are currently classified in the class with maximum classifications.

As shown in Fig. 7, the class potential $V_{\omega i}$ introduces an additional correction $\Delta f_{\omega i}$ of the assessment $f\omega_i$, formed by the ith RBE:

$$\Delta f_{\omega i} = k_v \cdot V_{\omega i}/D(X, C_i), \qquad (8)$$

where k_v is a weight coefficient, $D(X, C_i)$ is the Euclidean distance between the input vector X and the ith class center.

This correction displaces the probability density function $f_{\omega i}$ with a value, which is proportional to the current number of vectors of the class ω_i. The effect of the correction comes down to displacement of the boundary between two overlapping

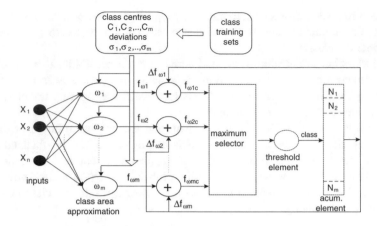

Fig. 7 Classifier architecture (CRBEP) which takes into consideration the class potentials

parts of class areas. The displacement depends on the ratio of accumulated number of vectors in each of the classes.

7 Hardware Description

The hardware system consists of the following main components: computer vision system (CVS) and spectrophotometer (4) (Fig. 8). The CVS includes two color CCD cameras (1) which give a possibility to form color images of investigated

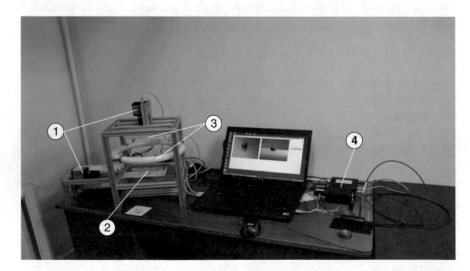

Fig. 8 INTECHN platform hardware components

object (2) in two planes (horizontal and vertical). The illuminant system (3) is used for direct object illumination. There is a possibility to include a second light source, which is located under the object. It is used to form images using transmitted light. INTECHN platform uses three kinds of spectrophotometers: QE65000 (4) (for grain sample analysis), NIRQuest 512 (for milk, dairy products, meat and meat products analysis) and CARY 100 (for fruits, vegetables and eggs analysis), as well as needed spectrometer accessories.

8 Results and Discussion

8.1 Training and Testing Sets

The developed procedures for grain sample quality assessment are validated, trained and tested with sets presented in Table 3.

The results from objects classification (non-grain impurities are included in testing sets) in shape and color classes are presented in Table 4.

The results from objects classification (non-grain impurities are included in testing sets) in normative classes of the selected classifiers on the basis of the three variants of data fusion are presented in Table 5.

The classification errors are calculated using the equations:

$$e_i = \frac{FN_i}{(TP_i + FN_i)} \tag{9}$$

e_i gives the relative part of objects from some class i, which are assigned incorrectly to other classes k = 1...N, where FN_i is the number of elements from the ith class

Table 3 Training and testing sets

Color classes recognition using CVS							
Classes	1cc	2cc	3cc	5cc	7cc	8cc	9cc
Training sets	10	10	12	15	18	19	
Testing sets	47	81	44	24	74	39	168
Object shape recognition							
Classes	1csh	2csh	3csh	4csh			
Training sets	120	135	135				
Testing sets	122	63	11	256			
Color classes recognition using spectra analysis							
Classes	1cc	2cc	3cc	5cc	7cc	8cc	9cc
Training sets	120	120	80	53	192	42	536
Testing sets	30	30	20	13	48	11	134

Table 4 Classification errors in shape and color classes

Color and shape class recognition						
	Shape class recognition		Color class recognition using CVS		Color class recognition using spectra analysis	
Selected classifiers	CRBEP		CDRBE		CDRBE Wavelet1 + PCA model	
Errors	Test. errors		Test. errors		Test. errors	
Class	g_i (%)	e_i (%)	g_i (%)	e_i (%)	g_i (%)	e_i (%)
1	39.5	5.7	8.9	8	0	15.2
2	62.0	69.8	8.3	4.9	0	5.1
3	87.5	77.8	0	15.9	0	0
4	21.9	43.1				
5			79.4	8.3	4.5	12.5
6						
7			35	13.5	19.5	5.4
8			0	5.3	0	10.3
9			23.2	68.4	12.0	6.6
	$e_0 = 35.6$ %		$e_0 = 30.6$ %		$e_0 = 7.3$ %	

Table 5 Classification errors in normative classes

Color and shape data fusion						
Fusion variant	Variant 1		Variant 2		Variant 3	
Selected classifiers	CDRBE-CRBEP		CDRBE-CRBEP		CRBEP-CDRBE Wavelet1 + PCA model	
Errors	Test. errors		Test. errors		Test. errors	
Class	g_i (%)	e_i (%)	g_i (%)	e_i (%)	g_i (%)	e_i (%)
1cst	7.7	28.2	4.2	1.7	2.7	6.8
2cst	27.9	32.2	17.2	14.4	0.8	12.7
3cst	12.7	0.8	6.2	9.1	8.4	0.4
	$e_0 = 15.3$ %		$e_0 = 8.6$ %		$e_0 = 5.3$ %	

classified incorrectly to other classes, TP_i is the number of correctly classified elements from the ith class;

$$g_i = \frac{FN_i}{(TP_i + FN_i)} \tag{10}$$

g_i gives the relative part of objects from other classes, which are assigned to class i, where FP_i is the number of elements from other classes assigned to the ith class;

$$e_0 = \frac{\sum_{i=1}^{N} FN_i}{\left(\sum_{i=1}^{N} TP_i + \sum_{i=1}^{N} FN_i\right)} \tag{11}$$

e_0 (classification error rate) gives the relative part of all incorrectly classified objects, were N is the number of classes.

1. The test results in shape class recognition show that the rate of objects from class 1csh assigned to other classes is comparatively small (4.9 %). On the other hand, the rate of objects assigned to this class which actually belong to other classes is sufficiently bigger (35.2 %). The rate of objects from class 3csh assigned to 2csh and 4csh is big too.
2. The classification error rate of objects from class 3csh (parts of kernels) is large. This is an expected result because it is impossible to define some standard shape for objects from this class. In many cases even a qualified expert will not recognize such objects if no color characteristics but only shape is taken into consideration. During the classifiers training models of broken kernels are created on the basis of whole kernels models and that is why the training sample classification error rates for classes 2csh and 3csh are small. This explains the big difference between training and testing classification results for these two classes.
3. The testing error in color class recognition using spectra analysis (7.3 %) is acceptable bearing in mind the specific investigation conditions and the diversity of grain sample elements.
4. The comparative analysis of the results obtained using different variants of classifier validation, training and testing confirms the effectiveness of the INTECHN classification strategy, classifiers, validation approach and data models. For example, if we use the three data models: PCA, Wavelet1 + PCA and Wavelet2 + PCA the training errors are 6.8, 6.3 and 10.3 % respectively using the CDRBE classifier. The INTECHN validation approach (when the non-grain impurities are included in validation procedure, but are excluded from training sets) decreases the testing error 3.8 times (from 27.6 to 7.3 %) in comparison with the traditional validation approach (when the non-grain impurities are simultaneously excluded or included in validation and training sets). The choice of an appropriate classifier for specific classification task has an influence over the classification accuracy too. For example, the training errors obtained using the CDRBE, CSRBE and CRBEP classifiers and PCA data model are 6.8, 72, and 7.3 % respectively.
5. The classification errors in color class recognition using spectra analysis are sufficiently smaller than the errors using image analysis. For example, the testing errors are 1.3 and 10 % respectively using the two approaches when the non-grain impurities are excluded from the validation and testing sets. When we include the non-grain impurities in validation and testing sets these two errors are 7.3 and 42 %. The big difference between the two errors can be explained by the fact that the object spectral characteristics contain not only information for

objects color characteristics, but for their surface texture too. Although typical for some grain groups color zones are found in a big part of non-grain impurities, the surface texture of these elements is sufficiently different from the typical for the grains.

6. Object classification in normative classes (1cst, 2cst and 3cst) includes complex assessment of color and shape characteristics of the investigated objects. For this purpose color data and shape data are fused. The data fusion procedure improves sufficiently the final classification results. The classification error rate e_0 in normative classes using CVS (Selected variant CDRBE–CRBEP) is 15.3 % when data fusion Variant 1 is used and 8.6 % when Variant 2 is used, while the errors of object color zones extraction and object shape recognition are 30.6 and 35.6 % respectively. When we use Variant 3 for classes' recognition the classification error rate e_0 decreases about 1.6 times in comparison with the better result obtained using CVS. This is due to the fact that the spectra analysis gives the best result in color classes' recognition.

9 Conclusions

The results from the investigation at this stage of the INTECHN project implementation concerning grain sample quality assessment using complex assessment on the basis of color image and spectra analyses can be summarized as follow:

1. The developed approaches, methods and tools for grain samples quality assessment based on the complex analysis of object color, object surface texture and object shape give an acceptable accuracy under specific experimental circumstances. The error rate $e_0 = 5.3$ % of the final categorization in the normative classes can be accepted as a good result at this stage of project implementation.
2. The data fusion procedure improves sufficiently the final classification results. The classification error rate e_0 using CVS is 15.3 % when Variant 1 is used and 8.6 % when Variant 2 is used, while the errors of object color zones extraction and object shape recognition are 30.6 and 35.6 % respectively.
3. The results obtained show that the choice of an appropriate procedure for fusion the results from color characteristics and objects shape analysis has a significant influence over the final classification accuracy. When we use the second algorithm (Variant 2) which is based on color or combined topology assessment the classification error rate decreases about 1.8 times compared to the first algorithm (Variant 1) in which color class assessment is based on the registration of the typical color zones combinations only. When we fuse the results from color classes recognition obtained on the basis of spectra analysis and shape classes recognition obtained on the basis of image analysis (Variant 3) the final classification accuracy is increased 2.9 and 1.6 times in comparison with Variant 1 and Variant 2 respectively.

Acknowledgements A big part of the analyses and results presented in this investigation are a part of implementation of the research project "Intelligent Technologies for Assessment of Quality and Safety of Food Agricultural Products", funded by the Bulgarian National Science Fund.

References

1. M.J. Aitkenhead, I.A. Dalgetty, C.E. Mullins, Weed and crop discrimination using image analysis and artificial intelligence methods. Comput. Electron. Agric. **39**, 157–171 (2003)
2. T. Brosnan, D.W. Sun, Inspection and grading of agricultural and food products by computer vision systems—a review. Comput. Electron. Agric. **36**, 193–213 (2003)
3. C. Damyanov, *Non Destructive Recognition of Quality in System for Food Products Automatic Sorting, Monography* (Academic publishing of the University of Food Technologies, Plovdiv, 2006)
4. F.E.E. Dowell, B. Maghirang, F. Xie, G.L. Lookhart, R.O. Pierce, B.W.S. Seabourn, R. Bean, J.D. Wilson, O.K. Chung, Predicting wheat quality characteristics and functionality using near-infrared spectroscopy. Cereal Chem. **83**(5), 529–536 (2006)
5. F. Dowell, T. Pearson, E. Maghirang, F. Xie, D. Wicklow, Reflectance and transmittance spectroscopy applied to detecting fumonisin in single corn kernels infected with fusarium verticillioides. Cereal Chem. **79**(2), 222–226 (2002)
6. A.D. Girolamo, V. Lippolis, E. Nordkvist, A. Visconti, Rapid and non invasive analysis of deoxynivalenol in durum and common wheat by Fourier-Transform infrared spectroscopy. Food Addit. Contam. **26**(6), 907–917 (2009)
7. H. Huang, H. Yu, H. Xu, Y. Ying, Near infrared spectroscopy for on/in-line monitoring of quality in foods and beverages: a review. J. Food Eng. **87**(3), 303–313 (2008)
8. G. Kos, H. Lohninger, R. Krska, development of a method for the determination of fusarium fungi on corn using mid-infrared spectroscopy with attenuated total reflection and chemometrics. Anal. Chem. **75**(5), 1211–1217 (2003)
9. K. Liao, M.R. Paulsen, J.F. Reid, B.C. Ni, B.E.P. Maghirang, Corn kernel breakage classification by machine vision using a neural network classifier. Trans. ASAE **36**(6), 1949–1953 (1993)
10. S. Mahesh, A. Manickavasagan, D.S. Jayas, J. Paliwal, N.D.G. White, Feasibility of near-infrared hyperspectral imaging to differentiate Canadian wheat classes. Biosyst. Eng. **101**(1), 50–57 (2008)
11. S. Mahesh, D.S. Jayas, J. Paliwal, N.D.G. White, Identification of western Canadian wheat classes at different moisture levels using near-infrared (NIR) hyperspectral imaging. CSBE Paper No. 08-196 (2008)
12. S. Mahesh, D.S. Jayas, J. Paliwal, N.D.G. White, Identification of wheat classes at different moisture levels using near-infrared hyperspectral images of bulk samples. Sens. Instrum. Food Qual. Saf. **2**(3–4), 1007–1015 (2010)
13. S. Majumdar, D.S. Jayas, Classification of cereal grains using machine vision, Part 3: texture models. Trans. ASAE **43**(6), 1681–1687 (2000)
14. S. Majumdar, D.S. Jayas, Classification of cereal grains using machine vision: color models. Trans. ASAE **43**(6), 1677–1680 (2000)
15. C. Miralbés, Prediction chemical composition and alveograph parameters on wheat by near-infrared transmittance spectroscopy. J. Agric. Food Chem. **51**(21), 6335–6339 (2003)
16. M.I. Mladenov, Grain Quality Analysis and Assessment. (Monography, Academic publishing of the University of Ruse, Ruse, 2011)
17. M. Mladenov, Pattern classification in a noisy environment. Inf. Technol. Control **1**(2011), 23–33 (2011)

18. M. Mladenov, M. Dejanov, Application of neural networks for seed germination assessment, in *Proceedings of the 9th WSEAS International Conference on Neural Networks* (2008), pp. 67–72

19. M. Mladenov, T. Draganova, M. Dejanov, S. Penchev, Assessment of grain quality features using image and spectra analyses, in *Proceedings of the International IFAC Workshop DYCAF* (Plovdiv, 2012), pp. 119–126

20. M. Mladenov, S. Penchev, M. Dejanov, M. Mustafa, Quality assessment of grain samples using colour image analysis, in *Proceedings of the 8th IASTED International Conference on Signal Processing, Pattern Recognition and Applications* (2011)

21. Y. Nakakimura, M. Vassileva, T. Stoyanchev, K. Nakai, R. Osawa, J. Kawano, R. Tsenkova, Extracellular metabolites play a dominant role in near-infrared spectroscopic quantification of bacteria at food-safety level concentrations. Anal. Methods **4**, 1389–1394 (2012)

22. H.F. Ng, W.F. Wilcke, Machine vision evaluation of corn kernel mechanical and mold damage. Trans. ASAE **41**(2), 415–420 (1998)

23. S. Ning, R. Ruan, L. Luo, X. Chen, P.L. Chen, R.K. Jones, Automation of a machine vision and neural network based system for determination of scabby rate of wheat samples. ASAE Annual International Meeting, Orlando, Florida (1988)

24. J. Paliwal, N.S. Visen, D.S. Jayas, Evaluation of neural network architectures for cereal grain classification using morphological features. J. Agric. Eng. Res. **79**(4), 361–370 (2001)

25. J. Paliwal, N.S. Visen, D.S. Jayas, Comparison of a neural network and non-parametric classifier for grain kernel identification. Biosyst. Eng. **85**(4), 405–413 (2003)

26. J. Paliwal, N.S. Visen, D.S. Jayas, N.D.G. White, Cereal grain and dockage identification using machine vision. Biosyst. Eng. **85**(1), 51–57 (2003)

27. M.R. Paulsen, L.O. Pordesimo, M. Singh, S.W. Mbuvi, B. Ye, Corn starch yield calibrations with near-infrared reflectance. Biosyst. Eng. **85**(4), 455–460 (2003)

28. M.R. Paulsen, M. Singh, Calibration of a near-infrared transmission grain analyzer for extractable starch in corn. Biosyst. Eng. **89**(1), 79–83 (2004)

29. T.C. Pearson, D.T. Wicklow, E.B. Maghirang, F. Xie, F.E. Dowell, Detecting aflatoxin in single corn kernels by transmittance and reflectance spectroscopy. Trans. Am. Soc. Agric. Eng. **44**, 1247–1254 (2001)

30. K.H.S. Peiris, M.O. Pumphrey, Y. Dong, E.B. Maghirang, W. Berzonsky, F.E. Dowell, Near-infrared spectroscopic method for identification of fusarium head blight damage and prediction of deoxynivalenol in single wheat kernels. Cereal Chem. **87**(6), 511–517 (2010)

31. R. Ruan, Y. Li, X. Lin, P. Chen, Non-destructive determination of deoxynivalenol levels in barley using nea-infrared spectroscopy. Appl. Eng. Agric. **18**(50), 549–553 (2002)

32. X14 30 S.H. Kutzbach, Determination of broken kernels in threshing process using image analysis and machine vision, in *Proceedings of 27th Symposium "Actual Tasks on Agricultural Engineering", Croatia* (1999)

33. R. Tsenkova, H. Meilina, S. Kuroki, D. Burns, Near infrared spectroscopy using short wavelengths and leave-one-cow-out cross-validation for quantification of somatic cells in milk. J. Near Infrared Spectro. **17**(6), 345–352 (2010)

34. Z.J. Wang, J.H. Wang, L.Y. Liu, W.J. Huang, C.J. Zhao, C.Z. Wang, Prediction of grain protein content in winter wheat (Triticum aestivum L.) using plant pigment ratio (PPR). Field Crops Res. **90**(2–3), 311–321 (2004)

35. I.J. Wesley, O. Larroque, B.G. Osborne, N. Azudin, H. Allen, J.H. Skerritt, Measurement of gliadin and glutenin content of flour by NIR spectroscopy. J. Cereal Sci. **34**, 125–133 (2001)

Intelligent Technical Fault Condition Diagnostics of Mill Fan

Mincho Hadjiski and Lyubka Doukovska

Abstract The mill fans (MF) are centrifugal fans of the simplest type with flat radial blades adapted for simultaneous operation both like fans and also like mills. The key variable that could be used for diagnostic purposes is vibration amplitude of MF corpse. However its mode values include a great deal of randomness. Therefore the application of deterministic dependencies with correcting coefficients is non-effective for MF predictive modeling. Standard statistical and probabilistic (Bayesian) approaches are also inapplicable to estimate MF vibration state due to non-stationarity, non-ergodicity and the significant noise level of the monitored vibrations. Adequate for the case methods of computational intelligence [fuzzy logic, neural networks and more general AI techniques—the precedents' method or machine learning (ML)] must be used. The present paper describes promising initial results on applying the Case-Based Reasoning (CBR) approach for intelligent diagnostic of the mill fan working capacity using its vibration state.

1 Introduction

The mill fans (MF) are a basic element of dust-preparing systems (DPS) of steam generators (SG) with direct breathing of coal dust in the furnace chamber. Such SG in Bulgaria is the ones in the Maritsa East 2 thermal power plant (TPP), in the Maritsa East 3 TPP and also in the Bobov Dol TPP.

The principal graph of a MF is shown in Fig. 1. After the row fuel bunker, the coal is dozed by a row fuel feeder 2, which is controlled by a cascade control system 3 as a task by the main power controller of the boiler-turbine unit. MF intake drying gases with a temperature 900–1000 °C from the upper side of the furnace

M. Hadjiski (✉)
University of Chemical Technology and Metallurgy, Sofia, Bulgaria
e-mail: zdravkah@abv.bg

L. Doukovska
Institute of Information and Communication Technologies—BAS, Sofia, Bulgaria

© Springer International Publishing Switzerland 2016
M. Hadjiski et al. (eds.), *Novel Applications of Intelligent Systems*,
Studies in Computational Intelligence 586, DOI 10.1007/978-3-319-14194-7_2

23

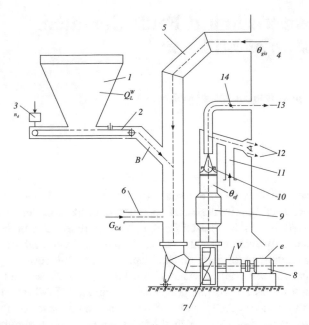

Fig. 1 Structure scheme, where *1*—Row fuel bunker, *2*—Row fuel feeder, *3*—Controller of row fuel feeder, *4*—Upper side of the furnace chamber, *5*—Gas intake shaft, *6*—Added cold air, *7*—Mill fan, *8*—Electric motor, *9*—Separator, *10*—Dust concentrator, *11*—Hot secondary air, *12*—Main burners, *13*—Discharge burner, *14*—Synchronized valves of discharge burners, θ_{af}—Temperature of air-fuel mixture, θ_{gis}—Temperature of intake drying gases, *V*—Vibration, *e*—Relative electric energy consumption, *B*—Throughput capacity of fuel, G_{CA}—Flow rate of added cold air, n_d—Position of discharge duct valve, Q_L^W—Low fuel caloricity of working mass

chamber 4 which in the drying shaft 5 are mixed with the raw coal and also with the additionally fed cold air 6. The MF 7 breaks and dries the raw coal actuated by the electric motor 8. The vibrations V of the MF are the basic topic of interest of the present research. The gravity-fed separator 9 guarantees the desired granularity of grinding of the output dust-air mixture which is directed towards the basic burners 12 via the dust concentrator 10 together with the added hot secondary air 11. The finest components (the discharges) are directed towards the discharge burners 13 which are controlled by synchronized valves 14.

The mill fans are a part of the equipment of power units that are most often repaired due to intensive erosion of the operating wheel blades in the process of grinding of low-calorific lignite coal from the Trayanovo 1 and Trayanovo 2 mines with high percentage of dust (28–45 % of the dry mass).

In spite of the constructive measures to use steels with high resistance index in critical parts of MF, their interval between two successive repairs equals to 2–2.5 months. The MF are of critical importance for the automation of the power-unit (PU) supply, so their technical state is an object of strict monitoring, repair and on-line duty according to the operative dispatcher time-table. The control range of mill fan is small, therefore the primary power control and especially the secondary power control is realized by stopping and starting some of the mill fan. Out of totally 8 grinding systems (two on a wall) in the 210 MW monoblocks in the Maritsa East 2 thermal power plant, 5–7 ones usually operate, 1 is ready and one/two of them are under repair. In this presentation the object of interest comprise fan MF with a horizontal axis of the operating wheel.

2 Mill Fan Technical Diagnostics

The mill fan has specificities that hamper significantly their diagnostics. The basic specifics are the following:

- The model-based MF diagnostics is insecure compared to e.g. rolling mills of coal with a vertical axis of rotation [1, 2] and ball mills [3–5]. The design of precise enough mathematical models are a sophisticated task due to the following items:
  - Basic thermo-mechanical and mode parameters are hampered or are impossible to be measured [6–8] (fuel consummation, granulometric composition, coefficient of grindability, coal quality).
  - Complex dependency of MF operation on a set of lots of parameters (coal composition, process hydrodynamics, exchange of heat and also of masses under changeable boundary conditions) [8–12].
  - Asymmetric wear of operative wheel blades, variable fan and grinding capacity between two successive repairs, Fig. 2 [6, 7, 13, 14].

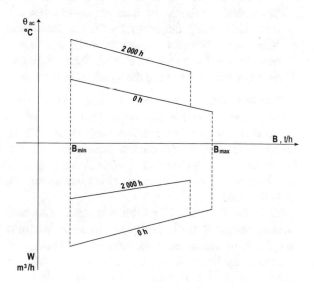

Fig. 2 Exploitation nomograms

Fig. 3 Relative power consummation for coal grinding

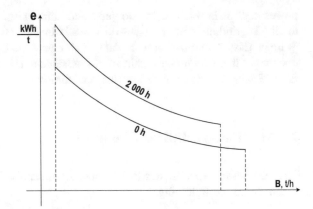

- The measurements of the necessary mode quantities for mill fan in the DCS or SCADA system are rather inaccurate due to the significant changeability of the conditions for measurements (wear, slagging, sensor pollution) and also due to the great amount of external disturbances (dust and humidity of fuel, imprecise dosing of coal, stochasticity of temperature of the intake oven gases due to non-stationarity of the torch position) [6, 7, 10–12].
- The period of starting work for MF after repairs must be set as a separate diagnostic problem because MF indicators for this period differ from the operational ones for the following 2000 h.
- The symptoms in a diagnostic problem are always indirect and they are of stochastic nature [changes of temperature of the dust-air mixture (DAM) at the separator output related to the average one for the DPS, the torch position (horizontal, vertical), the relative power consummation for grinding (Fig. 3), corrected rotating frequency for the raw fuel feeders (RFF)] [7, 10–12, 15].
- Vibrodiagnostics did not prove to be a serious diagnostic method for MF because regular measurements of their vibration state of the new DCS and SCADA turn out to be quite insufficient for a detailed diagnostics.

The MF vibration state may become a rather useful component of their diagnostics to determine their affiliation to some zone of efficiency: S_1—the normal one; S_2—partial damages; still possible exploitation with lowered mode parameters (e.g. loading) and measures for current maintenance (lubrication, jamming bolt joints of MF to the bearers, technological adjustments (angles of rotation of valves, jalousie); S_3—zone of serious damages, requiring immediate stopping at the first opportunity (stop the unit).

The mill fan state is multidimensional. The basic components are grinding productiveness B [t/h], fan productiveness W [m^3/h] and vibration state S_{MF}^V [amplitude of vibrations A (or velocity/acceleration of vibrations)].

Exemplary limits for the workability of the Maritsa East 2 thermal power plant monoblocks of 8 MW each are shown in Table 1.

Table 1 Exemplary limits for the workability

Component (S, dimension)	B (t/h)	W (m³/h)	A (mm)
S_1	55	200,000	6
S_2	48	180,000	7
S_3	<40	<140,000	>8

3 Mill Fan Vibrations Model

The results from the few references for research of the nature of MF vibrations [16, 17] and also of similar systems [18], our observations included and show that there are enough reasons to treat these vibrations as nonlinear.

Representing the model of vibrations of the mill fan corpse as an object with lumped parameters by the classical equation of linear vibrations [19],

$$\frac{d^2y}{dt^2} + 2\xi\frac{dy}{dt} + \omega_0^2 y = f_0\cos\omega t \tag{1}$$

does not reflect vibration behavior of MF.

Nonlinear vibrations may be represented in many ways [20]:

$$\frac{d^2y}{dt^2} + 2\xi\frac{dy}{dt} + F(y,t) = f(t) \tag{2}$$

The disturbances in the right side of Eq. (2) may be presented as a function of exciting mechanical disturbances (damaged bearings, unbalanceness due to wear, etc.) q_M and due to mode disturbances (loading, hydrodynamic instability) q_p, i.e.:

$$f(t) = f(q_M(t), q_P(t)) \tag{3}$$

Under certain assumptions without big limitations the function $f(t)$ may be treated as additive:

$$f(t) = f_1(q_M(t)) + f_2(q_P(t)) \tag{4}$$

Equation (2) becomes:

$$\frac{d^2y}{dt^2} + 2\xi\frac{dy}{dt} + F(y,t) = f_1(q_M(t)) + f_2(q_P(t)) \tag{5}$$

The principle of superposition is not applicable to nonlinear vibrations [20]. Therefore the exciting effect of the mode disturbances $q_P(t)$ must be eliminated or it must be rebuked substantially at the stage of analysis. The exciting disturbance $f_1(.)$ is of a deterministic nature and it is possible to be nonstationary if the fault develops (e.g. most often progressive wear leading to debalance). The mode disturbance $f_2(.)$ is of a cumulative nature (due to the co-effect of a variable loading, a change in the coal composition, hydrodynamic instability), it is stochastic. This may be used for processing of measured vibration signal to separate the effect from the mechanical excitement $f_1(.)$ of the observed vibrations.

The measurements for estimation of the MF vibration state—fault isolation (I) or diagnostics-in-depth (D)—require quite different frequency of discretization of the vibration signal $y(t)$ which is a solution of the nonlinear differential Eq. (5).

1. The fault isolation (case I) requires just the usage of discretization of the analog vibrosignal with a big quantization slice (e.g. $T_0 = 1$ [min], as it is accepted in DCS of Maritsa East 2 TPP). Signal $y(kT_0)$ is random, uncorrelated and for the purpose of our research it is characterized by two non-random characteristics.

Mathematical expectation:

$$Y_m(kT_0) = M[Y(kT_0)] \tag{6}$$

Mathematical deviance:

$$\sigma_Y(kT_0) = \sqrt{D[Y(kT_0)]} \tag{7}$$

Both values are functions of the discrete time $t = kT_0$. The signal with rare measurements $y(kT_0)$ (Fig. 4) is accepted in Maritsa East 2 TPP to avoid the excessive memory in the records' base for the process history, because the vibration signal is not at all used for diagnostic purposes and just for the purpose of additional control of the current technical state of the MF.

The performed by us research shows that the measured by a "regular" apparatus of DCS signal $y(kT_0)$ may be successfully used to detect faults (I) and it is an integral indicator for the vibration state of the MF. This may be realized only on condition that the effect of technological disturbances $f_2(q_p(kT_0))$ is reduced significantly in the summary vibrosignal $y(kT_0)$. In the case of MF, the problem of eliminating the technological disturbances is different and it is much more complicated than the predefined by some researchers of vibration problem for

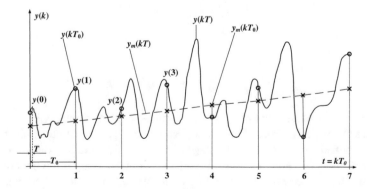

Fig. 4 The vibration signal (1 min sampling time)

de-noising of the signal [21, 22]. The basic reasons for this situation follow from the fact that there is no direct sensor information for the technological impacts and also because they are not only coordinate influences [the right side of Eq. (5)] but also parametric impacts influencing also parameters in the left part of Eq. (5). The uncertainty is substantially bigger, too. This imposed the elaboration of new methods for intelligent filtering of the technological impacts.

2. The in-depth vibrodiagnostics requires a spectral, wavelet and/or temporal-frequency analysis of the vibrosignal $y(t)$ (and its derivatives) and also to estimate the change(s) of its phase characteristics in the harmonic components. This is possible if the temporal analog signal of the vibrations ($y(t)$, $\frac{dy(t)}{dt}$ or $\frac{d^2y(t)}{dt^2}$) is quantized with a discretization interval satisfying the fundamental relation of Shannon-Kotelnikov. It requires in the case of MF a discretization interval T less than 1 [s] (Fig. 5). The obtained discrete signal $y(kT)$ is correlated and it may be used for spectral analyses together with the fast Fourier transform (FFT) for the purposes of a combined temporal-frequency analysis [22, 23], a wavelet analysis [19, 23] and other contemporary techniques for vibrodiagnostics [17, 19, 23]. It is possible to obtain a vibrosignal with a small discretization interval T via two approaches.

1. Apply frequent quantization in DCS with an interval of 300–500 [ms]. This approach requires adding a program code in the overall information system; this is a complex task and in the concrete case there is no agreement with the power plant managers. The drawback of the approach is the measurement just of the "overall vibration" which lacks information about the symptom in various places in the MF.

2. Include special diagnostic apparatus of the producing company that is applicable for routine vibrodiagnostics. The obtained signal $y(kT)$ then must be processed with original algorithms due to the peculiarities of MF vibrations and their multidimensionality [17, 19, 23].

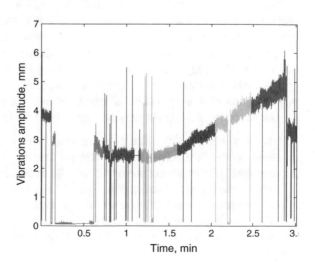

Fig. 5 Vibrations amplitude for entire period of observations

A great deal of the mode disturbances $q_p(t)$ over the vibration state S_{MF}^V of the MF are formed during the co-operation of the whole assembly of all operative MF. Such are: the drying gases temperature in the gas intake shaft, the torch position, the amount of the fuel that is synchronously distributed between the operative MF, the superheated steam. The differentiation between the individual part of the general impact $q_p^i(t)$ from the total one $q_p(t)$ is rather complex and it may be realized satisfactorily only applying intelligent techniques from computational intelligence [fuzzy logic, neural networks, support vector machines (SVM)] and processing knowledge (case-based reasoning (CBR), training) [4, 5, 10, 12, 16, 19].

4 Experimental Research

The experimental research is done in the national Maritsa East 2 thermal power plant. The plant has four double blocks with direct-current boilers 175 MW each and four monoblocks with drum boilers 210 MW each. The fuel for both types of blocks is one and the same: low-quality Bulgarian lignite coal from the Trayanovo 1 and Trayanovo 2 mines with caloricity of 1200–1600 kkal/kg (5000–6700 kkal/kg). The basic data are obtained from steam generator 6 of block 3. The steam generator has four MF. There is a decentralized control system (DCS) mounted over the steam generator: Experion PKS R301 Process of Honeywell. All used data are recorded in the Historian system of DCS. The duration of the observations is 8 months in 2010. Two types of data are used for the research—vibrosignals from MF and data about the basic mode parameters related to MF.

Figure 5 shows lines of discretized data about the MF vibration amplitude for a discretization time slice $T_0 = 1$ [min] and presents raw measurement data from the DCS for vibrations amplitude of the nearest to the mill fan motor bearing block.

The period of observations starts 7 days before mill fan rotor replacement; next are the data for the following 6 months exploitation period and at the end are measurements from the 7 days period after mill fan rotor replacement. The data are collected with 1 min interval. As can be seen from the figure, there are observed long stopping periods (when vibrations amplitude approaches zero) as well as different working regimes of the mill fan during its exploitation cycle.

The entire observations period is divided into sub-periods of 12 h each. Figure 6 present maximal density values of vibrations amplitude at each sub-period.

The following corollaries may be drawn from these data:

- The discretization time slice of the recorded data in the Historian files is $T_0 = 1$ [min]. These data belong to uncorrelated (due to the big values of T_0) random processes. Therefore according to the discussion above these temporal series may be used to isolate events in the MF vibration state but not for the detailed MF bearings' diagnostics because it is impossible to determine spectra of MF vibrations in successive time intervals due to the general non-stationarity of the process as a result of the wearing-out of the working wheel.

Fig. 6 Maximal density
values of vibrations amplitude

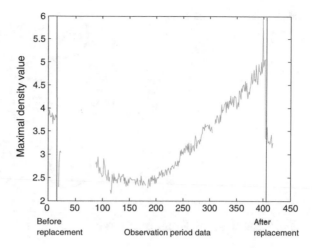

Before
replacement

Observation period data

After
replacement

- Vibrosignals demonstrate significant unstability of the MF oscillations due to series of random exciting powers q_p (Eq. 5)—a change in the fuel composition, non-homogeneous filling of sectors in the working wheel, hydrodynamic instability due to a change in the flow for the input and output cross sections of the MF. This instability is also due to often interrupts and load changes, provoked by corrections of the RFF rev regulator which is jointed in a cascade or due to a redistribution of the load of the main regulator for thermal block loading.

- Two exploitation periods are obviously formed—right after a repair (E_1) and after a normal exploitation (E_2). In the first case (E_1) there is a random change in the vibrations with different tendencies. In the second case (E_2) there is observed non-monotonous rise of vibrations due to the joint action of leading factors— erosive wearing-out of the blades leading to a debalance of the working wheel and a random combined influence of the enumerated above exciting the oscillations mode factors $(B_{MF}$—Throughput capacity of fuel, Q_L^W—Low fuel caloricity of working mass, θ_{af}—Temperature of air-fuel mixture, θ_{gis}—Temperature of intake drying gases, n_d—Position of discharge duct valve). These two periods must be analyzed apart.

- The root mean square deviation of the vibration amplitude σ_V is changed during the cycle of the working wheel from one repair to another (Fig. 6) and it is an additional symptom for an isolation of an abnormity and also for a forecast.

- Vibrosignals must be analyzed synchronously together with the extracts for the mode parameters $(\theta_{af}, \theta_{gis}, n_d)$ due to the high level of the noise in the causal-effective relations.

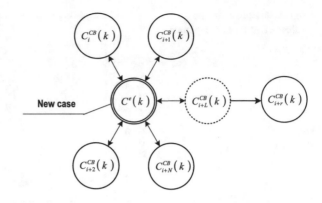

Fig. 7 Case-based reasoning model

5   Case-Based Reasoning Analysis

Due to the substantial ambiguity and variety of possible situations there is an additional procedure to specify diagnostic features and symptoms using Case-Based Reasoning, an approach enjoying an increasing popularity in the intelligent diagnostics [24–27] (Fig. 7).

In accordance with the settled tradition [24, 27] precedents are represented in the form "problem-solution":

$$C_i = (p_i, s_i). \tag{8}$$

The problem p_i is accepted with the structure "attribute-value":

$$p_i = (a_i, v_i). \tag{9}$$

The attributes vector includes the diagnostic features a_{ij} [28]:

$$a_i = (a_{i1}, a_{i2}, \ldots, a_{ir}). \tag{10}$$

The values $v_i(k)$ are related to the attributes a_i at the moment k and they are defined as creeping average analogical to formula (9). The averaging interval L is as it is known [19] an optimization problem and it is established experimentally.

$$v_i(k) = (v_{i1}(k), v_{i2}(k), \ldots, v_{ir}(k)). \tag{11}$$

The basic peculiarity of the application of the precedents' method in the case of diagnostics with mill fans with continuous degradation of their vibrational and technical state is the offer to use dynamic precedents depending on the time of observation k.

$$C_i = C_i(k). \tag{12}$$

Here k is the discrete time from the beginning of the mill fan campaign after the basic repair. So the source to form dynamic precedents are the archival records for all mill fans (eight of them) from the steam generator, operating with exploitation cyclic periodicity of 2000–2500 h. In this way each problem p_i (9) and the value v_i from the substantiation of the attribute a_i (10) are related to a fixed time moment k.

$$p_i = p_i(k)$$
$$v_i = v_i(k)$$
(13)

It was postulated that the formation of the cases will be performed via a time interval of $T_{CBR} = 2\ h$ according to the available data from DCS, soft sensing, mathematical modeling or an operator's decision.

The solution s_l in cases of diagnostics is presented in the form diagnostic state S—technical actions for technical support M:

$$s(k) = (S, M(k)).$$
(14)

According to the already made assumptions three diagnostic states are accepted:

- S_1—operable;
- S_2—conditionally allowed;
- S_3—unallowable.

Each of the diagnostic states S_j is related to a given discrete moment of time k and it also possesses a structure of the "attribute-value" type.

$$S_j(k) = (G, H(k)) \quad (j = 1, 2, 3).$$
(15)

The current state $S_{MB}(k)$ of a mill fan is related to some diagnostic state $S_j(k)$ using a classifier of the "comparison-with defined-thresholds" type based on the values $h_i(k)$ using a system of N rules R_i, for $(i = 1 \div N)$.

The set of attributes G consists of elements g_i:

$$G = (g_1, g_2, \ldots, g_m).$$
(16)

Each attribute g_i has a value $h_i(k)$ at the discrete moment of time k and it belongs to the vector $H(k)$.

$$H(k) = (h_1(k), h_2(k), \ldots, h_m(k)).$$
(17)

All values $h_i(k)$ are calculated as average with a procedure for creeping average for a given discrete time k using a formula analogous to (11).

The current state $S_{MB}(k)$ of a mill fan is related to some diagnostic state $S_j(k)$ using a classifier of the "comparison-with defined-thresholds" type based on the values $h_i(k)$ using a system of N rules R_i, for $(i = 1 \div N)$.

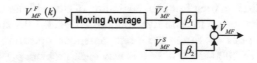

Fig. 8 Moving average structure scheme

$$R_i : IF h_1 < h_1^t u h_2 < h_2^t u \ldots h_5 < h_5^t \quad THEN \quad S \subset S_i. \tag{18}$$

In the present paper, a multistage procedure is accepted to estimate the mill fan vibrostate—S_{MB}^V, where the defined limits h_i^t are changed adaptively depending on the estimate of the root-mean-square value for the reduced noise in the registered vibrations (Fig. 8).

In the present paper a multistage procedure is accepted to estimate the mill fan vibration state—S_{MB}^V, where the defined limits h_i^t are changed adaptively depending on the estimate of the root-mean-square value for the reduced noise in the registered vibrations.

The actions for technical support M are presented as a multi-set [28]:

$$M = (M_1, M_2, M_3, M_4). \tag{19}$$

The components M_i ($i = 1 \div 4$) are subsets with the following components:

- M_1—change in the mode parameters in cases of conditionally allowed diagnostic state, e.g. with 3 elements.
- M_2—current repair, e.g. with 5 elements.
- M_3—replacing elements without big breaks of the mill fan operation, e.g. with 4 elements.
- M_4—stopping for repair, e.g. with 7 elements.

It is accepted in the paper that the basic part of the attributes in the problem section P and the solution S are presented by the simplest type of data: "number" and "symbol". Still for some attributes such representation by pairs "attribute-value" is incomplete and they [especially in the portion for the supporting activities M (19)] may include free text or they may contain links to other related external information. Part of this information may not be directly used in the CBR algorithm but it gives the operators an additional knowledge for secondary using of archived results from the mill fan exploitation. There originate certain difficulties to apply the approved procedure in order to follow the principle for local-global proximity [29–31] in cases of current diagnostics with a time attribute k from the beginning of the mill fan exploitation. The n-closest neighbors at the moment k are realized using the weighted proximity measure between two states I and J.

$$Sim(I, J) = \sum_{i=1}^{n} w_i sim_i(I_i, J_i) \quad \sum w_i = 1 \tag{20}$$

Here I_i and J_i denote the values and $w_i > 0$ is the weight of attribute a_i from (10). The symbol sim_i denotes a local proximity between the pair of diagnostic states I (k) and $J(k)$ at the moment k. It is accepted in the paper as a proximity measure sim_i to be normalized for the range $[0, 1]$; the Euclidean distance is used for the calculation in [31]. At the arrival of new data obtained through an interval of $T_{CBR} = 2$ h in cases of successive usage formula (19) is transformed in the following form.

$$Sim(p(k), p_i(k)) = \sum_{i=1}^{n} w_i sim_i(a(k), a_i(k)) \tag{21}$$

where $p(k)$ is the current "new" problem for the mill fan state, $p_i(k)$ is the existing ith precedent for the mill fan state at moment k; $a(k)$ and $a_i(k)$ are the respective attributes the values of which are presented by the vectors $v(k)$ and $v_i(k)$. The fraction $sim_i(a(k), a_i(k))$ representing the local proximity between the attributes a and a_i contains first of all knowledge for a specific domain (mill fan diagnostics) and the weight coefficients w_i reflect the relative meaning of these attributes over the determination of the common proximity between p and p_i. In the case greater weights are assigned to the fuel amount (recursively), the temperature of the aeromixture and the temperature of the gases in the gas intake shaft because they are measured following DCS data.

Generally the application of the method with precedents to determine the vibrational $S_{MB}^{V,\,CBR}(k)$ and the common S_{MB}^{CBR} diagnostic state is shown at Fig. 9. Block $CBR(n, k)$ is the generally accepted four-stage CBR procedure introduced in [29] and widely used during the following 15 years [19, 23, 30–33]. Estimates of the mill fan diagnostic state may be obtained with a discretization of $T_{CBR} = 2$ h, i.e. this is a slower approach that the one of intelligent filtration where the interval of discretization may be 20 [min].

Independently on the presented significant difficulties during the determination of the vibration state of MF—S_{MB}^V, it is advisable to include it as an important component in the assessment of the overall technical state of mill fan. The assessment of the mill fan vibration state is a complex problem due to the exceptionally big uncertainty in the measurements which follows from the temporally re-covered changes of multimode factors.

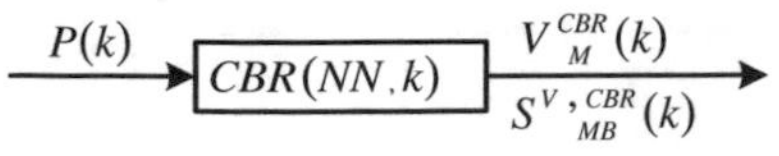

$$P(k) \longrightarrow \boxed{CBR(NN, k)} \begin{array}{l} V_M^{CBR}(k) \\ \hline S^{V,\,CBR}_{MB}(k) \end{array} \longrightarrow$$

Fig. 9 CBR diagnostic block

The mill fan vibration state (S_{MB}^V) is a valuable integral indicator for its working capacity. The determination and the usage of mill fan vibration state indicators are realistic and profitable for the operative staff because vibrosensors are obligatory for contemporary decentralized control DCS systems.

Figures 10, 11, 12, and 13 contain some of the processed vibrations data aimed at excluding outliers. Only data that are around maximal density value of each 12 h sub-period are left. This is done due to mentioned high non-stationary nature of

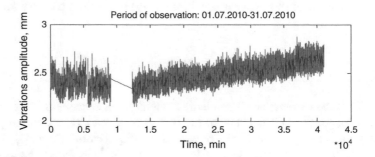

Fig. 10 Processed data for the period of observation 01.07.2010–31.07.2010

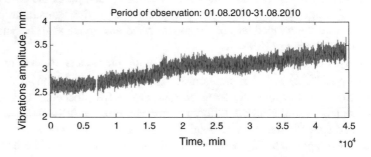

Fig. 11 Processed data for the period of observation 01.08.2010–31.08.2010

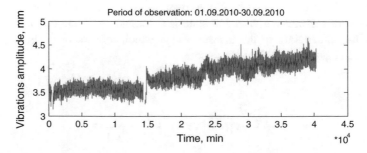

Fig. 12 Processed data for the period of observation 01.09.2010–30.09.2010

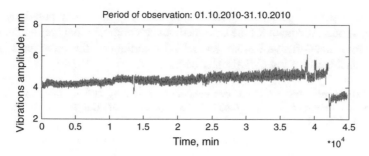

Fig. 13 Processed data for the period of observation 01.10.2010–31.10.2010

these data. Another purpose of this processing was to exclude stopping periods from our investigation.

6 Conclusion

The vibrosignals may be successfully used as a substantial additional symptom for isolation and diagnostics of mill fan system which is not done at present. The assessment of the MF vibration state is a complex problem due to the exceptionally big uncertainty in the measurements which follows from the temporally re-covered changes of multimode factors. The MF vibration state S_{MB}^V is a valuable integral indicator for its efficiency.

Independently on the discussed significant difficulties during the definition of the vibration state of MF—S_{MB}^V, it is advisable to include it as an important component in the assessment of the overall technical state of MF.

The records from existing DCS (or SCADA) do not allow a frequency (e.g. spectral) analysis but they may be used to isolate and to forecast defects at the stage of "isolation". An increase of the discretization frequency (e.g. 100 times) is unacceptable for DCS due to unforeseen in the computational resources and software.

Separate "short" records as long as 2–3 days (2000–4000 point of data) must be processed due to the non-stationarity of the vibrosignal that are enough as a volume for a representative statistical analysis of the time series at an admissibly low non-stationary change of the mill fan vibrostate.

Standard statistical and probabilistic (Bayesian) approaches for diagnostics are inapplicable to estimate MF vibration state due to non-stationarity, non-ergodicity and the significant noise level of the monitored vibrations. Promising results are obtained only using computational intelligence methods (fuzzy logic, neural and neuro-fuzzy networks).

In this paper are presented promising results only using computational intelligence methods. Adequate for the case methods of computational intelligence [fuzzy logic, neural networks and more general AI techniques—the precedents' method (CBR), machine learning (ML)] must be used.

Acknowledgements This work has been partially supported by FP7 grant AComIn № 316087 and partially supported by the National Science Fund of Bulgaria, under the Project No. DVU-10-0267/10.

References

1. G. Fan, N. Rees, *Modeling of Vertical Spindle Mills in Coal Fired Power Plants* (1994)
2. N. Rees, G. Fan, Modeling and control of pulverized fuel coal mills. IEE Power Energy Series **43** (2003)
3. J. Wei, J. Wang, S. Guo, Mathematical modeling and condition monitoring of power plant tube-ball mill systems, in *Proceedings of ACC* (St. Louis, USA, 2009)
4. P. Zachariades, J. Wei, J. Wang, Development of a tube-ball coal mill mathematical model using particle swarm optimization, in *Proceedings of the World Congress of Engineering* (London, 2008)
5. H. Cao, L. Jia, G. Si, Y. Zhang, In improved simulated annealing for ball-mill pulverizing system optimization of thermal power plant, in *Advances in Information Technology and Industry Application, LNEE 136* (Springer, 2012)
6. B. Bonev, T. Totev, J. Artakov, M. Nikolov, Diagnostic of coal dust preparation systems with milling fans, in *Proceedings of the 3rd International Conference "New Trends in Automation of Energetic Processes'98"* (Zlin, Czech Republic, 1998)
7. T. Totev, B. Bonev, J. Artakov, *Systems for Determination of the Quality of Coal, Burning in the Steam Generator P-62 in TPP "Maritza East Energetics"*, No. 1–2, 1995 (in Bulgarian)
8. S. Batov, B. Bonev, T. Totev, *Problems and Decisions of the Use of Bulgarian Low-Rank Lignite Coal for Electric Power Generation, Proc.* (STC, Ochrid, 1995)
9. L. Hadjiski, S. Dukovski, M. Hadjiski, R. Kassing, Genetic algorithm application to boiler-mill-fan system robust optimal control, in *Symposium Algarve* (Portugal, 1996)
10. M. Hadjiski, M. Nikolov, S. Dukovski, G. Drianovski, E. Tamnishki, Low-Rank Coal Fired Boilers Monitoring by Applying Hybrid Models, in *Proceedings of the 8th IEEE Mediterranean Conference on Control and Automation MED'2000* (Patras, Greece, 2000a)
11. M. Hadjiski, V. Totev, Hybrid modeling of milling fan of steam generators in TPP. Autom. Inf. **4** (2000) (in Bulgarian)
12. M. Hadjiski, V. Totev, R. Yusupov, Softsensing-based flame position estimation in steam boiler combustion chamber, in *Proceedings of International Workshop on "Distributed Computer and Communication Networks"* (Sofia, Bulgaria, 2005)
13. N. Klepiko, V. Abidennikov, A. Volkov et al., Dust preparation systems with mill-fans for the boilers of power units, Teploenergetika, No. 9, 2008 (in Russian)
14. M. Hadjiski, S. Dukovski, Adaptive Control of 210 MW Boiler Milling System—Comparative Analysis of Several Approaches, in *International Symposium on Adaptive Control/Signal Processing* (Budapest, Hungary, 1995)
15. M. Hadjiski, V. Petkov, E. Mihailov, A Software environment for approximate model design of a low-calorithic coal combustion in power plant boilers, in *DAAP Symposium on "Modelling and Optimization of Pollutant Reduced Industrial Furnaces"* (Sofia, Bulgaria, 2000b)
16. M. Tabaszewski, Forecasting of residual time of milling fans by means of neural networks. *Diagnostyka'3* **39** (2006)

17. C. Cempel, M. Tabaszewski, Singular spectrum analysis as a smoothing method of load variability. *Diagnostyka'4* **56** (2010)
18. N. Zhang, *Dynamic Characteristic of Flow Induced Vibration in a Rotor-Seal System* (IOS Press, Amsterdam, 2010)
19. J. Norbicz, J. Koscielny, Z. Kowalczuk, W. Cholewa (eds.), *Fault Diagnosis Models, Artificial Intelligence, Applications* (Springer, Berlin, 2004)
20. F. Moon, *Chaotic and Fractal Dynamics* (Wiley, New York, 2008)
21. D. Donoho, Denoising by soft-thresholding. IEEE Trans. Inf. Theor. **41** (1995)
22. S. Krishnan, R. Rangayyan, Denoising knee joint vibration signals using adaptive time-frequency representation, in *Canadian Conference on Electrical and Computer Engineering*, vol. 3 (1999)
23. G. Vachtsevanos, F. Lewis, M. Roemer, A. Hess, B. Wu, *Intelligent Fault Diagnosis and Prognosis for Engineering Systems* (Wiley, New York, 2006)
24. J. Recèo-Garcia, B. Diaz-Agudo, A. Sanches-Ruiz, P. Gonzales-Calero, *Lessons Learned in the Development of a CBR Framework*, Expert Update, vol. 10, No. 1 (2010)
25. A. Eremeev, P. Varshavskiy, Case-based reasoning method for real-time expert diagnostics systems. Int. J. Inf. Theor. Appl., vol. 15, 2008
26. I. Rasovska, B. Chebel-Mollero, N. Zerhouni, A, Case elaboration methodology for a diagnostic and repair help system based on case-based reasoning, in *Proceedings of AAAI* (2007). www.aaai.org
27. S. Pal, S. Shin, *Foundation of Soft Case-Based Reasoning* (Wiley, New York, 2004)
28. M. Hadjiski, L. Doukovska, CBR approach for technical diagnostics of mill fan system. Comptes rendus de l'Academie bulgare des Sciences, ISSN 1310–1331 **66**(1), 93–100 (2013)
29. J. Kolonder, *Case-Based Reasoning* (Morgan Kaufmann, Burlington, 1993)
30. M. Richter, On the notion of similarity in CBR, in *Mathematical and Statistical Methods in Artificial Intelligence*, ed. by G. del la Riccia, R. Kusse, R. Viertl (Springer, Berlin, 1995)
31. A. Aamodt, E. Plaza, Case-based reasoning: foundational issues, methodological variations, and system approaches. AI Commun. **7**(1) (1994)
32. L. Fan, K. Boshnakov, Fuzzy logic based dissolved oxygen control for SBR wastewater treatment process, in *Proceedings of the 8th World Congress on Intelligent Control and Automation* July 6–9 2010 (Jinan, China, 2010), pp. 4142–4144
33. L. Fan, Z. Yu, K. Boshnakov, Adaptive backstepping, based terminal sliding mode control for DC-DC converter, in *International Conference on Computer Application and System Modeling (ICCASM)* (Tayuan, 2010), pp. V-323–V9-327

Abstraction of State-Action Space Utilizing Properties of the Body and Environment

Analysis of Policy Obtained by Three-Dimensional Snake-Like Robot Operating on Rubble

Kazuyuki Ito, So Kuroe and Toshiharu Kobayashi

Abstract We focused on the autonomous control of a three-dimensional snake-like robot that moves on rubble. To realize an autonomous controller, we employed reinforcement learning. However, applying reinforcement learning in a conventional framework to a robot with many degrees of freedom and moving in a complex environment is difficult. There are three problems: state explosion, lack of reproducibility, and lack of generality. To solve these problems, we previously proposed abstracting the state-action space by utilizing the universal properties of the body and environment. The effectiveness of the proposed framework was demonstrated experimentally. Unfortunately, analysis of the obtained policy was lacking. In the present study, we analyzed the obtained policy (i.e., Q-values of Q-learning) to determine the mechanism for abstraction of the state-action space and confirmed that the three problems were solved.

1 Introduction

The current progress in the mechanical performance of robots is remarkable; practical applications to complex environments such as rescue operations are expected. Snake-like robots are one candidate for such applications. In this study,

K. Ito (✉)
Hosei University, 3-7-2, Kajino-cho, Koganei-shi, Tokyo 1848584, Japan
e-mail: ito@hosei.ac.jp

S. Kuroe
Yokogawa Electric Corporation, 2-9-32, Naka-chou, Musashino-shi,
Tokyo 180006, Japan

T. Kobayashi
Mitsubishi Chemical Corporation, 1-1 Marunouchi 1-chome, Chiyoda-ku,
Tokyo 100-8251, Japan

© Springer International Publishing Switzerland 2016
M. Hadjiski et al. (eds.), *Novel Applications of Intelligent Systems*,
Studies in Computational Intelligence 586, DOI 10.1007/978-3-319-14194-7_3

we considered the autonomous control of a three-dimensional snake-like robot that moves on rubble.

The development of autonomous controllers for robots with many degrees of freedom has been delayed, and operating snake-like robots autonomously in unpredictable complex environments is difficult.

To realize autonomous control, the controller requires a function to learn effective behaviors for a given environment.

Application of reinforcement learning to autonomous robots is a significant challenge [1–16]. Reinforcement learning can allow a robot to learn effective behavior for a given task in a given environment: it can learn by trial and error with no supervisor required. This will make it possible to operate a robot in an unpredictable environment. Various experiments have demonstrated the effectiveness of combining reinforcement learning with a robot in an unpredictable environment [4–6, 14, 15]. However, since complex environments have various problems, these conventional successful applications were restricted to robots in simple environments.

There are three main problems with learning in complex environments such as rubble:

- *State explosion.* Complex environments have a great deal of information that needs to be learned. In addition, the robot usually has many degrees of freedom to operate in the complex environment, so the amount of information for the robot is also huge. As a result, the size of the state-action space increases exponentially, and reinforcement learning becomes impossible.
- *Lack of reproducibility.* In a complex environment like rubble, the same situation is almost never encountered twice. Thus, in the conventional framework of reinforcement learning, the learning machine faces a new state every time. Therefore, the obtained policy does not improve, and the learning process often becomes impossible to complete.
- *Lack of generality.* The obtained policy of reinforcement learning usually has no generality and depends on the learning environment. Therefore, even if an effective policy is obtained in a certain environment, the policy will not work effectively in other environments. Therefore, additional learning is required every time the environment changes. With real rubble, every environment is different. Therefore, a policy learned through conventional reinforcement learning is practically worthless.

In conventional works, various approaches have been proposed to reduce the size of the state-action space. Combining reinforcement learning with neural networks [6–8], fuzzy control [9–12], stochastic approach [13], or genetic algorithms [14, 15] is an effective approach. However, some of these approaches require prior knowledge to design the controller; in addition, for complex environments, the computations to reduce the state-action space incur a significant computational cost. Therefore, applying the proposed methods to complex and controlled objects such as a snake-like robot moving on rubble is difficult. Real-time learning using a real robot is also very difficult.

In the real world, higher organisms learn in real time by trial and error, and acquired learning has general applicability. The reason behind such organisms' learning ability is still debated. In embodied cognitive science [16] and ecological psychology [17], the body and environment are considered to play important roles in realizing adaptive behavior. However, the roles of the body and environment in learning have not been adequately discussed.

In previous works, we focused on the roles of the body and environment and proposed generalizing the state-action space by using the properties of the body and the actual environment. We demonstrated that reinforcement learning can be applied to a snake-like robot by designing its body so that it can adapt to the target environment. We also demonstrated that autonomy and versatility are not lost even if we use the properties of the environment [18–20]. Unfortunately, these previous works focused on realized behavior; analysis of the obtained policy was left to future work.

The present study was based on the experiments of our previous work [20]: we analyzed the obtained policy (Q-values of Q-learning) to determine the details of abstraction of the state-action space and confirmed that the three problems were solved.

2 Task and Environment

We considered a three-dimensional snake-like robot that moves on rubble. Figure 1 shows the target environment. To build the artificial rubble environment, we used 1036 wooden blocks placed randomly. The size of the environment was 333 cm × 252 cm. The robot was to move towards a light source in the environment. Table 1 lists the size and number of wooden blocks. The aim of the task was

Fig. 1 Example of artificial rubble

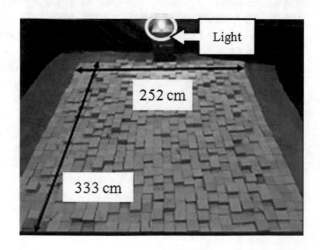

Table 1 Size and number of wooden blocks

Size of the blocks	1 cm 9 cm 9 cm	3 cm 9 cm 9 cm	5 cm 9 cm 9 cm	7 cm 9 cm 9 cm
Number of the blocks	496	180	180	180

for the robot to learn effective behavior to move towards the light source in this three-dimensional environment.

To confirm the generality of the obtained policy, we considered two cases:

- *Case 1*. The robot learns in a three-dimensional environment and obtains an effective control law. We then reposition the wooden blocks randomly and apply the obtained control law to the new environment without relearning.
- *Case 2*. The robot learns in a simple horizontal plane to acquire the effective control law. The obtained control law is then applied in a three-dimensional environment consisting of wooden blocks.

3 Proposed Method

3.1 Importance of Design Based on Environment

As noted in Sect. 1, applying conventional reinforcement learning to a snake-like robot that moves on rubble is very difficult. However, higher organisms in the real world can learn in real time by trial and error and apply the acquired policies generally.

In our previous works, we focused on real-world properties to explain the difference between the learning mechanism of robots and that of higher organisms, and we proposed a new framework in which the robot body uses real-world properties to solve these problems [18–20]. We summarize the previous discussions as follows. In typical reinforcement learning, real-world properties are ignored and idealized, and the learning process is mathematically described using an algorithm. We believe this to be the cause of the problems described in Sect. 1. In contrast, organisms have bodies suited to their environment and they adapt themselves by using the properties of that environment. In biology, this is called an ecological niche. Therefore, higher organisms may utilize the properties of their environment to improve their learning efficiency. Furthermore, even if a body depends on its environment, its versatility and autonomy are not lost. This is because an organism lives in the environment and can always use its properties. Thus, the problems can be solved not by improving a universal learning algorithm but by improving the body to utilize the properties of the environment. Details are presented in [19].

In our previous work [20], we employed this framework and designed a three-dimensional snake-like robot that can learn on rubble within a reasonable time limit. The present paper is based on the experiment of the previous work; we analyzed the obtained policy (Q-values of Q-learning) to determine the details of the abstracted policy.

3.2 General Framework

Figure 2 shows the proposed general framework. The robot consists of a generalization module and learning module. The remarkable feature of the proposed framework is that the generalization module is realized by the body instead of computers. The body is designed to utilize the properties of the real world such as the dynamics of the body, and necessary calculations for generalization are conducted using real-world properties.

The abstracted information is passed from the learning module to a computer, the learning process is executed on a computer, and the selected action is passed on to the generalization module. The generalization module embodies the abstracted action to the complex movement of the robot. Generalization of the information processing in the learning module ensures that the problem described in Sect. 1 is solved. In the following subsection, we present the design of the body for the snake-like robot based on this framework.

3.3 Hardware Design of Body

We developed a three-dimensional snake-like robot based on the proposed framework. Figure 3 shows the mechanism of the proposed robot. The head of the robot is an acute-angled triangle, and it has small free wheels to reduce friction with obstacles (Fig. 4). Every link has two active wheels and one crawler that move the robot (Fig. 5). We did not use any actuators for joints; instead, each link is connected by a rubber pole and passively moves along three directions (yaw, roll, and pitch)

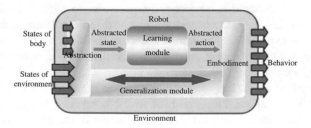

Fig. 2 Abstraction of state-action of spaces

Fig. 3 Robot outline

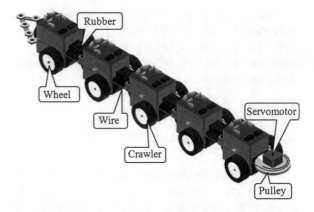

Fig. 4 Head of robot

Fig. 5 Crawler and wheel of robot

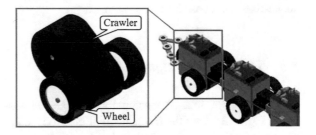

(Fig. 6). We did not employ force sensors on the body or angle sensors for the joints. The use of this passive mechanism implies that each link adapts to the environment without any sensors or controllers.

However, because the passive mechanism cannot control the direction of motion of the robot, we employed two wires and an actuator to control the length of the wires. As shown in Fig. 7, the wires are embedded in the robot, and their length is controlled by a rear-end servomotor. By reeling the wire on one side using the servomotor (Fig. 8), the shape of the robot is curved, and the direction of movement changes (Fig. 9).

When the robot encounters obstacles, some free joints are moved by a reactive force (Fig. 10). The affected links pull the wire on one side, and the wire pulls the other links to compensate for the change in direction of the robot. When the robot avoids the obstacle, the rubber pulls the affected links and returns them to their

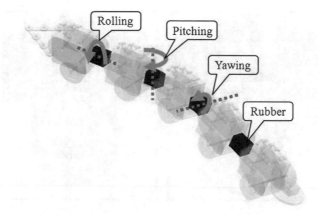

Fig. 6 Robot outline (yaw, roll and pitch)

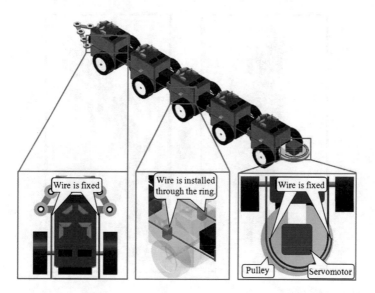

Fig. 7 Wire-pulling mechanism

initial straight shape. If more than one obstacle hits the body, the resultant force of the reactive forces determines which links move. As a result of the wire constraints and dynamics of the rubber, the robot moves as shown in Fig. 11. If the length of wires on both sides is equal and the obstacles are set uniformly, the expected result is straight movement while the obstacles are avoided (Fig. 11a). If the length of wire on the left side is shorter than the wire on the right side, the expected movement is a turn to the left (Fig. 11b).

The movement of each joint is realized passively in this mechanism. The joint to be moved and its extent of motion are determined by real-world dynamics.

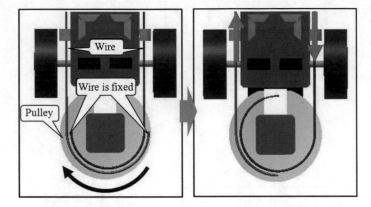

Fig. 8 Active pulley movement

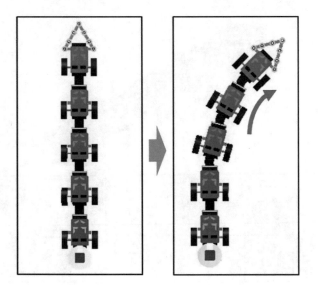

Fig. 9 Turning mechanism

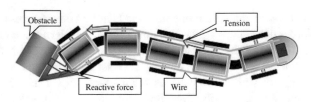

Fig. 10 Passive mechanism

Fig. 11 Robot behavior. **a**
Move forward. **b** Turn left

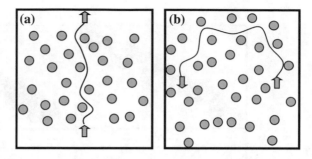

The decision of whether to overcome or avoid an obstacle is also determined from
real-world dynamics (Fig. 12a, b). Consequently, the microbehavior of avoiding or
overcoming obstacles is passively controlled, and the macrobehavior of moving to
the expected direction is controlled by a single actuator—i.e., the tail motor.

In summary, we can reduce the state of the body to a single dimension: i.e., the
differences between lengths of wires. Furthermore, we can reduce the action space
to a single dimension by restricting the action to moving the tail motor and
changing the wire lengths.

3.4 Hardware Design for Sensing

We employed many CdS cells and used the equilibrium point of the output of the
CdS cells for the direction. Figure 13 shows a CdS cell, a module of CdS cells, and
their layout in the robot. CdS cells are electrical cells that convert light intensity to
electrical resistance. We developed one module by placing six CdS cells as shown
in Fig. 13. Because the CdS cells have directional characteristics, we can obtain the
direction of the light source. In this layout, the perceivable range of the horizontal
direction is about 180° and that of the vertical direction is about 90°. We embedded
the module into every link in the robot, as shown in Fig. 13c, and calculated (1) in
parallel using an analog electrical circuit; the obtained result was used as the
direction, where x is the horizontal direction ($0 < x < 3$) and Rij is the electrical
resistance of the CdS cell of the ith link and jth position. Equation (1) describes the
center of gravity of the light intensity; the vertical direction is ignored, and only the
horizontal direction is obtained. When light is incident in front of the robot, x is 1.5.
If x is smaller than 1.5, this implies that the light is on the right side of the robot. If
x is greater than 1.5, this implies that the light is on the left side of the robot.

$$x = \frac{\sum_{i=1}^{4} (1/R_{i2}) + \sum_{i=1}^{4} (1/R_{i3}) + 2\{\sum_{i=1}^{4} (1/R_{i4}) + \sum_{i=1}^{4} (1/R_{i5})\} + 3\sum_{i=1}^{4} (1/R_{i6})}{\sum_{i=1}^{4} \sum_{j=1}^{6} (1/R_{ij})}$$

$$(1)$$

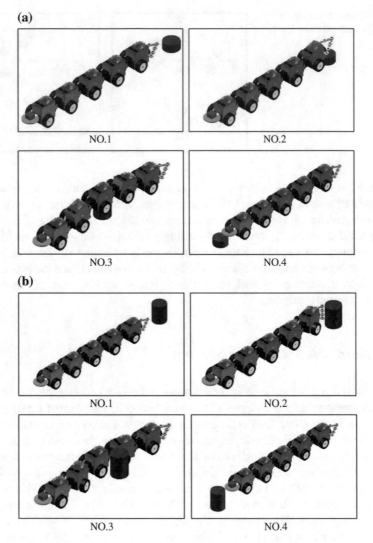

Fig. 12 Switching behaviors: **a** Robot overcomes obstacles lower than the head triangle. **b** Robot avoids obstacles higher than the head triangle

3.5 Developed Robot

Figure 14 and Table 2 show the developed robot and list its specifications, respectively. The passive joints are made from a rubber material of different dimensions (height, width, and depth). By changing the rubber material's dimensions, we adjust the elasticity in each direction (yaw, roll, and pitch). The different

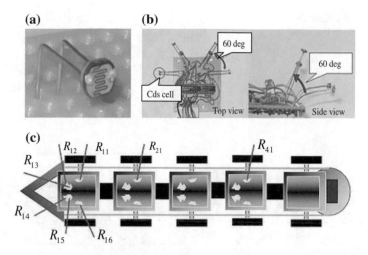

Fig. 13 Sensing system using CdS cells. **a** Cds cell. **b** Module of Cds cell. **c** Layout of the modules

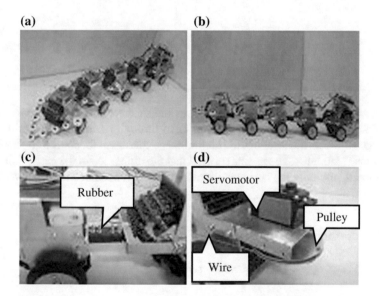

Fig. 14 Snake-like robot. **a** Robot. **b** Robot (lateral view). **c** Joint. **d** Pulley

Table 2 Specifications of snake-like robot

Length (mm)	800
Width (mm)	160
Height (mm)	140
Weight (g)	1200

Fig. 15 Light direction and
output of receiving circuit

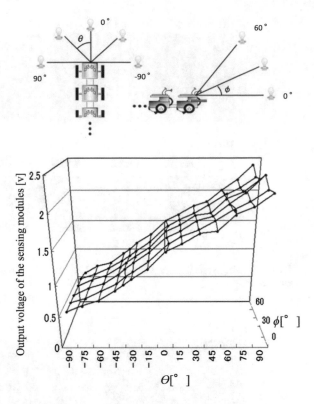

dimensions are very important to realizing the desired adaptive behavior. In the
present study, the dimensions were fine-tuned through preliminary experiments.

Figure 15 shows the results of the preliminary experiment for confirming the ability
of the sensor modules. The results showed that the difference in the vertical direction
was ignored, and only the horizontal direction was obtained as the output voltage.

4 Experiment

4.1 Setting of Reinforcement Learning

We employed simple Q-learning because our aim was to demonstrate that the body
itself can abstract the state-action space without improving the learning algorithm.
Equation (2) describes Q-learning:

$$Q(s,a) \leftarrow (1-\alpha)Q(s,a) + \alpha\left\{r(s,a) + \gamma \max_{d} Q(s',a')\right\} \tag{2}$$

where s is the state, a is the action, r is the reward, α is the learning rate, and γ is the discount rate.

We set α as 0.2 and γ as 0.5. The action was selected using the ε-greedy method, and the probability of random selection was 0.1. The duration of one trial was 50 s, and the calculations were performed using (2) every 2.5 s. Table 3 lists the states of the light direction. The resulting values are the outputs of (1). Table 4 lists the states of the body. These values represent the angle of the tail motor. As shown in Tables 3 and 4, there were two state space dimensions and a total of 25 states. Table 5 lists the actions executed by the robot. The actions consisted of turning the tail motor, which pulled the wires.

When the robot was going toward the light source (i.e., state number of the light direction was two and state number of the angle of the tail motor was two), a reward of 100 is given. When the robot loses the light source (light source goes out of the perceivable range), a negative reward of -100 is given. The trial breaks, and the next trial starts from the initial position.

4.2 Experiment

As noted in Sect. 2, we considered two cases. Figures 16, 17, 18 and 19 show the results of Case 1, and Figs. 20, 21, 22 and 23 show the results of Case 2.

In Case 1, the learning process was conducted in the environment shown in Fig. 16, and the effective behavior was obtained after 50 trials. Figure 16 shows the behavior of the fiftieth trial, and Fig. 17 shows its state transition diagram.

Table 3 Light direction

State	Voltage (V)
0	0.75–1.00
1	1.00–1.25
2	1.25–1.75
3	1.75–2.00
4	2.00–2.25

Table 4 Angle of the tail motor

State	0	1	2	3	4
Degree (°)	−50	−25	0	25	50

Table 5 Action

Action	Motion
0	Turn tail motor −25°
1	Hold tail motor
2	Turn tail motor −25°

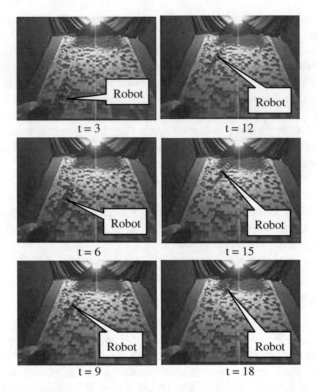

Fig. 16 Learned behavior in Case 1

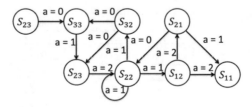

Fig. 17 State transition diagram of Fig. 16

We then repositioned the wooden blocks and applied the obtained policy to the new environment without relearning. Figure 18 shows the realized behavior in the new environment, and Fig. 19 shows its state transition diagram.

In Case 2, the learning process was conducted on the simple horizontal plane shown in Fig. 20, and the effective behavior was obtained after 50 trials. Figure 20 shows the behavior of the fiftieth trial, and Fig. 21 shows its state transition diagram.

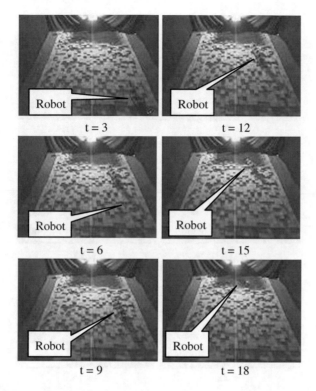

Fig. 18 Realized behavior in Case 1

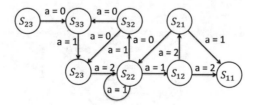

Fig. 19 State transition diagram of Fig. 18

We then applied the obtained policy to the three-dimensional environment without relearning. Figure 22 shows the realized behavior in the three-dimensional environment, and Fig. 23 shows the corresponding state transition diagram.

These results showed that the effective behavior to move toward the light source was obtained within a reasonable time limit, and the acquired policy can be applied to a different environment without relearning. By appropriately designing the body,

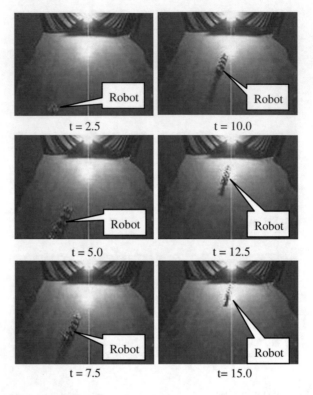

Fig. 20 Learned behavior in Case 2

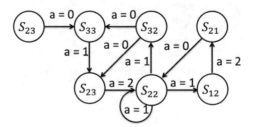

Fig. 21 State transition diagram of Fig. 20

the policy learned in a simple environment can be applied to a complex environment without additional learning. This is the remarkable feature of the proposed framework.

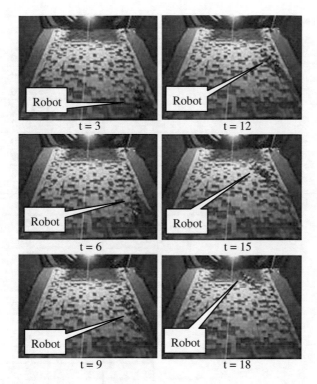

Fig. 22 Realized behavior in Case 2

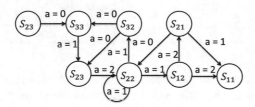

Fig. 23 State transition diagram of Fig. 22

5 Discussion

Table 6 shows the obtained Q-values for Case 1, and Table 7 shows the obtained values for Case 2. The maximum value for each state is highlighted in boldface type.

In general, the robot behaves optimally by conducting the action with the highest Q-value in each state. Thus, the positions of the highest Q-value in each state are important to analyzing the obtained policy. If the Q-value is negative, the action in that state causes a failure. In these experiments, the initial Q-values were 0.00.

Table 6 Obtained Q-values for Case 1

State		Action		
Light source	Motor	0	1	2
0	0	0.00	0.00	0.00
	1	−13.00	0.00	0.00
	2	−50.00	−50.00	**1.67**
	3	−50.00	−75.00	−20.90
	4	−50.00	0.00	−14.20
1	0	0.00	0.00	0.00
	1	−18.70	**73.39**	−50.00
	2	−5.68	−50.40	**68.12**
	3	0.00	−50.00	**10.65**
	4	0.00	0.00	0.00
2	0	0.00	0.00	0.00
	1	**99.89**	0.00	−22.50
	2	0.00	**99.98**	5.49
	3	0.00	4.80	**97.65**
	4	0.00	0.00	0.00
3	0	0.00	−22.50	−50.00
	1	**16.13**	−46.40	−50.00
	2	**61.73**	0.00	−10.60
	3	0.00	**78.75**	1.27
	4	0.00	0.00	0.00
4	0	−50.00	−50.00	−50.00
	1	−9.90	−50.00	−75.00
	2	**22.78**	−50.00	−75.00
	3	0.00	0.00	0.00
	4	0.00	0.00	0.00

Thus, if the Q-value was still 0.00 after the experiments, this means that the robot's situation (combination of state and action) never occurred during the learning process.

First, we confirmed the obtained policy presented in Table 6. When the light source state was 0, the light source was on the extreme right of the sensible range of the robot. Therefore, the robot was likely to lose the light source unless it turned right. Thus, almost all the Q-values in the row where the light source state was 0 were negative. The Q-value was only positive when the state of the motor was 2 (i.e., angle of motor was 0) and action was 2 (i.e., turn the motor to the right): that is, the robot turned right and moved to the light.

Next, we confirm the case when the light source state was 1. In this case, the light source was on the right of the robot. Action 1 (hold the motor) had the highest Q-value in the row where the state of the motor was 1 (i.e., angle of motor was −25°). This can mean that as the motor had already turned to the right, the robot can turn to the right by continuing to hold the motor.

Table 7 Obtained values for Case 2

State	Action			
Light source	Motor	0	1	2
0	0	0.00	0.00	0.00
	1	0.00	0.00	0.00
	2	−50.00	−50.00	**75.84**
	3	0.00	0.00	0.00
	4	−50.00	0.00	0.00
1	0	0.00	0.00	0.00
	1	19.23	**85.71**	0.00
	2	−50.00	0.00	**82.81**
	3	−50.00	0.00	**53.57**
	4	0.00	0.00	0.00
2	0	0.00	0.00	0.00
	1	**88.60**	36.56	0.00
	2	−25.00	**86.30**	2.50
	3	0.00	0.00	**98.93**
3	4	0.00	0.00	0.00
	0	−27.50	0.00	0.00
	1	−3.93	0.00	−22.50
	2	**68.86**	−7.12	−32.60
	3	0.00	83.65	0.00
	4	0.00	0.00	0.00
4	0	−50.00	−50.00	−75.00
	1	−80.60	−75.00	−87.50
	2	**53.63**	−50.00	−75.00
	3	0.00	0.00	0.00
	4	0.00	0.00	0.00

When the state of the motor was 1 (i.e., angle of motor was 0) or 2 (i.e., angle of motor was +25°), as the motor had not yet turned to the right, the action to turn the motor to the right (action 2) had the highest Q-value.

Next, we confirmed the case when the light source state was 2. In this case, the light source was ahead of the robot. Actions that moved the robot straight ahead had the highest Q-values. When the state of the motor was 1 (i.e., angle of motor was −25°) or 3 (i.e., angle of motor was +25°), the action to turn the motor to the straight position (actions 0 or 2) had the highest Q-value. If the state of the motor was 2 (i.e., angle of motor was 0°), action 1 (hold the motor) had the highest value.

The cases where the light source state was 3 and 4 were similar to the cases where the light source state was 0 and 1, respectively: they were symmetrical.

Based on these results, we confirmed that a policy that moves the robot to the light source was obtained.

Note that the policy does not contain information on the complex environment and body. The policy comprises only simple abstracted information such as the direction to the light source and moving direction of the robot. In other words, even in a complex environment, the machine learns only simple rules such as "if a light source is on the right (left) then turn right (left)" and "if a light source is located in front then go forward." Moreover, the robot can move to the desired direction on rubble with these simple rules despite the many degrees of freedom of its body.

These results showed that the abstraction and embodiment of the proposed framework (Fig. 2) were successfully realized.

Next, we discuss the obtained policy on the flat plane. Table 7 shows that the positions with the highest Q-value were the same as those in Table 6 except for one cell (light state = 3, motor state = 1, action = 0). This means that the policies of Tables 7 and 6 were the same despite the robot learning in different environments. As shown in Fig. 22, the policy learned on the flat plane was effective in the rubble environment, and the realized state transition (Fig. 23) was similar to that on the flat plane (Fig. 21). The mechanism that enables the effective behavior can be explained by the identity of the obtained policy. Due to the abstraction mechanism of the body, the different environments are observed as the same simple environment. Thus, the obtained policy is very simple and independent of the environment. The simple policy is embodied by the body to complex behavior that is effective for a given environment.

These results confirmed that differences in environments were compensated for, and the problems of a lack of generality and lack of reproducibility were solved. Note that the mechanism of abstraction and embodiment was realized by the body utilizing the universal properties of the environments. As this process was conducted simultaneously outside the computer, the problem of real-time processing did not occur even when the environment was very complex. The size of the state action space was very small (only 25 × 3) owing to this abstraction mechanism, so real-time learning is also possible. The results confirmed that the state explosion problem was solved.

In summary, we confirmed that the problems of conventional reinforcement learning were solved by abstraction of the state-action space, which was realized by utilizing the universal properties of the body and environment.

6 Conclusion

This study focused on the abstraction of state-action space in reinforcement learning and application to a three-dimensional snake-like robot. The idea was to design the body of the snake-like robot to utilize the universal properties of the environment for abstraction. In this framework, the abstraction process is conducted outside the computer by the interaction of the body with the environment.

We conducted the learning process in real environments and analyzed the obtained policies. The results showed that the abstraction process was successfully realized, and the same simple policy was obtained in different environments. Owing to this abstraction mechanism, different rubble environments were observed as a simple horizontal plane. Therefore, the lack of reproducibility problem was solved.

As the complex environments were observed as a simple environment, the size of the state-action space was greatly reduced, so the state explosion problem was solved. A simple policy was embodied to complex behavior that was effective for a given environment; the simple policy was applicable to various environments. Thus, the lack of generality problem was solved.

From these results, we conclude that abstracting state-action space by utilizing the properties of the body and environment is very effective for robot learning in real complex environments. Our future work will involve applying the proposed framework to other practical tasks such as rescue operations.

References

1. R.S. Sutton, *Reinforcement Learning: An Introduction* (The MIT Press, Massachusetts, 1988)
2. L.P. Kealbling, M.L. Littman, Reinforcement learning: a survey. J. Artif. Intell. Res. **4**, 237–285 (1996)
3. C.J.H. Watkins, P. Dayan, Technical note Q-learning. Mach. Learn. **8**, 279–292 (1992)
4. M. Asada, Y. Katoh, M. Ogino, K. Hosoda, A humanoid approaches to the goal: reinforcement learning based on rhythmic walking parameters, in *Proceedings of the International Symposium of RoboCup2003* (2003, CD-ROM)
5. H. Kimura, T. Yamashita, S. Kobayashi, Reinforcement learning of walking behavior for a four-legged robot, in *Proceedings of 40th IEEE Conference on Decision and Control* (2001), pp. 411–416
6. K. Doya, H. Kimura, M. Kawato, Neural mechanism of learning and control. IEEE Control Syst. Mag. **21**(4), 42–44 (2001)
7. C. Anderson, Z. Hong, Reinforcement learning with modular neural networks for control, in *Proceedings of NNACIP'94: IEEE International Workshop on Neural Networks* Applied to *Control and Image Processing* (1994)
8. W. Sun, X. Wang, Y. Cheng, Reinforcement learning method for continuous state space based on dynamic neural network, in *Proceedings of the 7th World Congress on Intelligent Control and Automation* (2008), pp. 750–754
9. D. Gu, H. Hu, Reinforcement learning of fuzzy logic controller for quadruped walking robots. Presented at Proceedings of 15th IFAC World Congress, Barcelona, Spain, 21–26 July 2002
10. H.R. Berenji, Fuzzy learning for generalization of reinforcement learning, in *Proceedings of the Fifth IEEE International Conference on Fuzzy Systems*, vol. 3 (1996), pp. 2208–2214
11. W.M. Hinojosa, S. Nefti, U. Kaymak, Systems control with generalized probabilistic fuzzy-reinforcement learning. IEEE Trans. Fuzzy Syst. **19**(1), 51–64 (2011)
12. A. Likas, Reinforcement learning using the stochastic fuzzy min-max neural network. Neural Process. Lett. **13**, 213–220 (2001)
13. T. Inamura, M. Inada, H. Inoue, Integration model of learning mechanism and dialogue strategy based on stochastic experience representation using Bayesian network, in *Proceedings of the 2000 IEEE International Workshop on Robot and Human Interactive Communication* (2000), pp. 247–252
14. S. Ushio, M. Svinin, K. Ueda, S. Hosoe, An evolutionary approach to decentralized reinforcement learning for walking robots, in *Proceedings of the 6th International Symposium on Artificial Life and Robotics* (2001), pp. 176–179
15. K. Ito, F. Matsuno, Reinforcement learning for redundant robot: solution of state explosion problem in real world, in *Proceedings of ROBIO'05 Workshop on Biomimetic Robotics and Biomimetic Control* (2005), pp. 36–41
16. R. Pfeifer, *Understand Intell*, new edn. (The MIT Press, Massachusetts, 2001)

17. J.J. Gibson, *The Ecological Approach to Visual Perception* (Lawrence Erlbaum Associates, Hillsdale, NJ, 1987)
18. K. Ito, A. Takayama, T. Kobayashi, Hardware design of autonomous snake-like robot for reinforcement learning based on environment: discussion of versatility on different tasks, in*Proceedings of the 2009 IEEE/RSJ International Conference on Intelligent Robots and Systems* (2009), pp. 2622–2627
19. K. Ito, Y. Fukumori, A. Takayama, Autonomous control of real snake-like robot using reinforcement learning: abstraction of state-action space using properties of real world, in *Proceedings of the International Conference on Intelligent Sensors, Sensor Networks and Information Processing* (2007), pp. 389–394
20. K. Ito, S. Kuroe, T. Kobayashi, Abstraction of state-action space utilizing properties of the body and the environment: application to a three-dimensional snake-like robot that operates on rubble, in *Proceedings of IEEE International Conference on Intelligent Systems* (2012), pp. 114–120

Trajectory Control of Manipulators Using an Adaptive Parametric Type-2 Fuzzy CMAC Friction and Disturbance Compensator

Kostadin Shiev, Sevil Ahmed, Nikola Shakev and Andon V. Topalov

Abstract Friction and disturbances have an important influence on the robot manipulator dynamics. They are highly nonlinear terms that cannot be easily modeled. In this investigation an incrementally tuned parametric type-2 fuzzy cerebellar model articulation controller (P-T2FCMAC) neural network is proposed for compensation of friction and disturbance effects during the trajectory tracking control of rigid robot manipulators. CMAC networks have been widely applied to problems involving modeling and control of complex dynamical systems because of their computational simplicity, fast learning and good generalization capability. The integration of fuzzy logic systems and CMAC networks into fuzzy CMAC structures helps to improve their function approximation accuracy in terms of the CMAC weighting coefficients. Type-2 fuzzy logic systems are an area of growing interest over the last years since they are able to model uncertainties and to perform under noisy conditions in a better way than type-1 fuzzy systems. The proposed intelligent compensator makes use of a newly developed stable variable structure systems theory-based learning algorithm that can tune on-line the parameters of the membership functions and the weights in the fourth and fifth layer of the P-T2FCMAC network. Simulation results from the trajectory tracking control of two degrees of freedom RR planar robot manipulator using feedback linearization techniques and the proposed adaptive P-T2FCMAC neural compensator have shown that the joint positions are well controlled under wide variation of operation conditions and existing uncertainties.

K. Shiev (✉) · S. Ahmed · N. Shakev · A.V. Topalov
Department of Control Systems, Technical University of Sofia, Campus Plovdiv,
4000 Plovdiv, Bulgaria
e-mail: k.shiev@gmail.com

S. Ahmed
e-mail: sevil.ahmed@tu-plovdiv.bg

N. Shakev
e-mail: shakev@tu-plovdiv.bg

A.V. Topalov
e-mail: topalov@tu-plovdiv.bg

© Springer International Publishing Switzerland 2016
M. Hadjiski et al. (eds.), *Novel Applications of Intelligent Systems*,
Studies in Computational Intelligence 586, DOI 10.1007/978-3-319-14194-7_4

63

1 Introduction

The performance requirements in today's manufacturing environments concerning the speed and accuracy of motion have increased considerably. Through the years different robot manipulator control schemes have been proposed to provide an improved trajectory tracking performance. Many of them can be regarded as special cases of the class of computed-torque controllers [19]. Computed-torque (CT) control approach is used to derive conveniently very effective robot controllers providing an exact knowledge of the robot model is available. At the same time, it is a special application of the feedback linearization of nonlinear systems, which has gained popularity in modern systems theory [9, 12]. For the rigid-link arms the two control approaches are equivalent.

It has recently been recognized that the accuracy of the conventional control approaches in high-speed applications is greatly affected by the presence of unmodeled dynamics caused by friction forces, vibration, backlash and uncertainties in the parameters describing the dynamic properties of the grasp load (e.g., unknown moments of inertia) [18]. To compensate for the existing uncertainties and unmodeled dynamics, many researchers have proposed adaptive and learning strategies for the control of rigid robot manipulators that ensure asymptotic trajectory tracking [11]. The development of effective adaptive controllers represents an important step toward high speed/precision robotic applications.

In this investigation an on-line tuned parametric type-2 fuzzy cerebellar model articulation controller (P-T2FCMAC) network is proposed for compensation of friction and disturbance effects during the trajectory tracking control of rigid robot manipulators. Cerebellar model articulation controller (CMAC) networks have been widely applied for identification and control of complex nonlinear dynamical systems with unmodeled nonlinearities and uncertain disturbances due to their computational simplicity, fast learning and good generalization capabilities [13]. They are non-fully connected perceptron-like associative memories that are able to compute nonlinear functions by referring to look-up tables over the domain of interest [2, 3]. The contents of these memory locations are denoted as weights, and the network output is calculated as a linear combination of the weights that are addressed by the activated inputs [15]. The basic idea implemented in the CMAC networks is to store learned data into overlapping memory regions in a way that the data can be easily recalled and the storage space is minimized [16]. The mechanism of storing the weight information is similar to that used in the cerebellum of human beings.

The integration of fuzzy logic and CMAC networks into fuzzy CMAC structures helps improving the function approximation accuracy in terms of the CMAC weighting coefficients and could offer a viable approach for modelling, nonlinear identification and control of nonlinear dynamic systems that are inherently uncertain and imprecise [5]. Type-2 fuzzy logic systems (T2FLSs) are an area of growing interest over the last years. They have been credited to be more powerful than their type-1 counterparts in compensating for even higher degrees of both structured and unstructured uncertainties [10, 18, 21] and are particularly suitable for modeling time–variant systems with unknown time-varying dynamics. The membership functions of the type-2 fuzzy sets are three-dimensional and include a footprint of uncertainty (FOU). The new third dimension provides additional degree of freedom that makes it possible to directly model uncertainties [22]. It has been also shown that T2FLSs have better noise reduction property when compared to type-1 systems [14].

Stability of the on-line learning neuro-fuzzy structures is an important issue [27]. Existing investigations have been split over three main research directions. Lyapunov approach can be directly implemented to obtain robust training algorithms [20]. Input-to-state stability is another elegant approach [27]. A third way to design robust learning schemes is to utilize the variable structure systems (VSS) theory [8]. Such systems exhibit the robustness and invariance properties inherited from variable structure control technique while still maintaining good approximation capability and flexibility.

A new stable incremental learning algorithm for P-T2FCMAC neural networks, based on the variable structure systems theory principles, is proposed in this investigation. It can tune the parameters of the membership functions and the weights in the network fourth and fifth layer. The algorithm is also able to adapt the existing relation between the lower and the upper membership functions of the type-2 fuzzy system. The latter helps managing non-uniform uncertainties.

The paper is organized as follows. Section 2 presents an overview of the manipulator dynamics and CT control. The implemented P-T2FCMAC neural network and the developed sliding mode control (SMC) theory-based on-line learning algorithm are presented in Sect. 3. The proposed trajectory tracking control scheme using the adaptive P-T2FCMAC friction and disturbance compensator is introduced in Sect. 4. Simulation results are shown in Sect. 5. Finally, concluding remarks are given in Sect. 6.

2 Manipulator Dynamics and Computed-Torque Control

2.1 The Dynamical Model

The dynamical equation of a general n degrees-of-freedom rigid-robot manipulator in joint space, without the actuator dynamics, can be written in the following form [18]:

$$M(q)\ddot{q} + V_m(q, \dot{q})\dot{q} + F(\dot{q}) + G(q) + \tau_d = \tau \qquad (1)$$

where $M(q) \in R^{n \times n}$ is the inertia matrix, $V_m(q, \dot{q}) \in R^n$ the Coriolis/centripetal vector, $G(q) \in R^n$ the gravity, $F(\dot{q}) \in R^n$ the viscous and the dynamic friction, $\tau_d(t) \in R^n$ a disturbance, $\tau(t) \in R^n$ the control input, and $q \in R^n$ the joint variable vector.

2.2   Computed-Torque Control

Computed-torque method is a basic robot manipulator control approach and at the same time represents a special application of feedback linearization of nonlinear systems. It provides a uniform framework for applying different control laws and one way to classify the existing robot control schemes is to divide them into "computed-torque-like" and "noncomputed-torque-like" [18]. CT-like controls appear in robust control, adaptive control, and learning control. CT takes into account the dynamic coupling between the manipulator links and results in a completely decoupled error dynamics equation. To ensure trajectory tracking by the joint variable let us denote the desired manipulator trajectory by $q_d(t)$ and define an output or tracking error as

$$e(t) = q_d(t) - q(t) \tag{2}$$

The tracking error dynamics of the system (1) can be written as

$$\frac{d}{dt} \begin{bmatrix} e \\ \dot{e} \end{bmatrix} = \begin{bmatrix} 0 & I \\ 0 & 0 \end{bmatrix} \begin{bmatrix} e \\ \dot{e} \end{bmatrix} + \begin{bmatrix} 0 \\ I \end{bmatrix} u \tag{3}$$

where

$$u = \ddot{q}_d + M^{-1}(q)(V_m(q, \dot{q})\dot{q} + G(q) + F(\dot{q}) + \tau_d - \tau)) \tag{4}$$

At the first design stage the method includes determination of the control law u that stabilizes the linear system (3). After that the appropriate control input τ is calculated

$$\tau = M(q)(\ddot{q}_d - u) + V_m(q, \dot{q})\dot{q} + G(q) + F(\dot{q}) + \tau_d \tag{5}$$

3   The Adaptive Parametric Type-2 Fuzzy CMAC Compensator

The P-T2FCMAC network combines the concepts of CMAC networks and type-2 Takagi-Sugeno-Kang (TSK) fuzzy neural networks implementing fuzzy if-then rule base with first order output functions.

3.1   Fundamentals of CMAC Networks

The association mechanism of CMAC networks, generally presented as an input-output mapping [21, 22], results in network structure shown on Fig. 1.

In CMAC, each input dimension is quantized and divided into discrete elements, and several elements are combined to form a block [7]. A possible quantization of the input space for a two dimensional CMAC can be seen on Fig. 1.

Consider two input variables x_1 and x_2 belonging to the interval $[-1, 1]$. Each input is divided into 4 discrete elements (sub-regions) as follows: $(-1, -0.5)$, $(-0.5, 0)$, $(0, 0.5)$ and $(0.5, 1)$. These sub-regions are further grouped into 2 blocks (regions) in each receptive layer respectively. The generated blocks for x_1 are labeled with $A_1, B_1, C_1, D_1, E_1, F_1$. Same labeling convention is also applied to the other input—x_2. Combining the blocks from the same layer of quantization of the three variables produces multi-dimensional receptive fields called hyper-cubes. In the current example the hyper-cubes corresponding to the first quantization layer of the three inputs are labeled as A_1A_2, A_1B_2, B_1A_2, B_1B_2. The next two layers of quantization generate receptive fields in a similar way. Thus, the mapping vector a^T could be defined as:

$$a^T = [h_1 \ h_2 \ h_3 \ \ldots h_{N_h}] \tag{6}$$

where N_h is the number of hyper-cubes ($N_h = 12$), which are labeled as in (7).

The mapping vector a^T is called also association vector. In the memory association, each vector element is activated or not according to the state of the corresponding input—x_1 or x_2. For instance, the input sequence $[x_1 x_2]^T = [0.25 \ -0.25]^T$ activates the following fields: A_1A_2, D_1D_2 and F_1E_2. This means that the

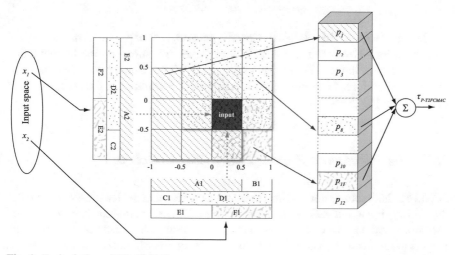

Fig. 1 Basic design of 2D CMAC

corresponding hyper-cubes h_1, h_8 and h_{11} will be set to 1. The other elements of the vector a^T are set to 0 (non-active).

$$
\begin{array}{lll}
\boxed{h_1 = A_1 A_2} & h_5 = C_1 C_2 & h_9 = E_1 E_2 \\
h_2 = A_1 B_2 & h_6 = C_1 D_2 & h_{10} = E_1 F_2 \\
h_3 = B_1 A_2 & h_7 = D_1 C_2 & \boxed{h_{11} = F_1 E_2} \\
h_4 = B_1 B_2 & \boxed{h_8 = D_1 D_2} & h_{12} = F_1 F_2
\end{array}
\tag{7}
$$

Based on the obtained memory, the CMAC associates each activated receptive field to a corresponding physical memory in order to aggregate the CMAC output:

$$
y_{CMAC} = \sum_{i=1}^{Nh} h_i p_i = a^T P \tag{8}
$$

where p_i denotes the ith element of the CMAC weight vector P, which presents the physical memory of the network (Fig. 1).

As a result of the processes of quantization, receptive field composition and correlation mapping, only few receptive fields will be activated and will contribute to the network output [7].

The number of mapping vectors corresponds to the size of the CMAC hyperspace that is calculated as $N_e = n^I$, where n is the number of the sub-regions that depends on the selected resolution and I is the size of the input space (the number of input variables). For the given example of 2D CMAC $N_e = 4^2 = 16$. This number covers all possible variances of memory activation by combining the blocks from each receptive layer according the input vector's values. Thus, the associative memory space A of the CMAC network can be presented as:

$$
A = [a_1 \quad a_2 \quad \ldots \quad a_{Ne}]^T \tag{9}
$$

Nevertheless, in the classical CMAC networks, the physical memories are stimulated by a binary value vector. This imposes limitations on the interpretation of the network inputs [7]. Fuzzy logic can be implemented to overcome this problem. Therefore, fuzzy inference mechanism is integrated in the receptive field composition [6, 7, 26].

3.2 The Parametric Type-2 Fuzzy CMAC Network

A CMAC internal memory is implemented as a bit memory location connected to the binary input activation functions. Fuzzy CMAC networks use real-valued memories, and each of them corresponds to the membership grade of the input with respect to the fuzzy activation functions [18]. In parametric fuzzy CMAC networks

(P-FCMAC) these constant valued memories are replaced by parametric equations, in terms of the network inputs, and the membership grades of these inputs with respect to fuzzy activation functions.

The architecture of the proposed in this chapter parametric type-2 fuzzy CMAC neural network is based on the one introduced by Ahmed et al. [1]. Therefore, an interval type-2 Takagi-Sugeno-Kang (TSK) fuzzy if-then rule base is implemented as in (10):

$$R_r: if \ x_1 \ is \ \tilde{A}_{1j} \ and \ x_2 \ is \ \tilde{A}_{2j} \ and \ \dots \ and \ x_i \ is \ \tilde{A}_{ij}$$
$$then \quad f_r = a_r^T P \tag{10}$$

where x_i, $(i = 1, 2, \dots I)$ is the input variable; $\tilde{A}_{ij}$ is the jth interval type-2 membership function for ith input variable; f_r, $(r = 1, 2, \dots N)$ is the fuzzy inference output function; a_r^T is the association vector for rth rule corresponding to the activated receptive fields; $P = [p_1 \ p_2 \ p_3 \dots p_{N_h}]$ is the weighting column vector of the P-T2FCMAC network represented by first order TSK functions $p_j = \sum_{i=1}^{I} b_{ij} x_i + b_{0j}$, where $j = 1, 2, \dots N_h$.

The general multi-input single-output (MISO) parametric type-2 fuzzy CMAC (P-T2FCMAC) architecture consists of five layers: input quantization layer, fuzzyfication layer, fuzzy association layer, fuzzy post-association layer and output layer (see Fig. 2).

Layer 1. *Input quantization*: There are no differences between the input layers of a CMAC and a P-T2FCMAC network. In this layer each input variable x_i from the I-dimensional input space of the network is quantized into n discrete sub-regions according to a defined resolution. A possible quantization of the input (sensor) space for two dimensional CMAC network is presented in Fig. 1.

Layer 2. *Fuzzification layer*: The quantized in layer 1 input space is fuzzified here by predefined fuzzy sets with corresponding membership functions. A type-2 fuzzification of the input space of the fuzzy CMAC network is implemented in this investigation. Interval type-2 fuzzy sets (IT2-FSs), presented by type-2 Gaussian membership functions with uncertain standard deviation (Fig. 3), have been used into the second layer of the P-T2FCMAC architecture.

The lower and upper membership functions for type-2 Gaussian membership functions with uncertain deviation are expressed by:

$$\bar{\mu}_{ij}(x_i) = \exp\left(-\frac{1}{2}\frac{(x_i - c_{ij})^2}{\bar{\sigma}_{ij}^2}\right) \tag{11}$$

$$\underline{\mu}_{ij}(x_i) = \exp\left(-\frac{1}{2}\frac{(x_i - c_{ij})^2}{\underline{\sigma}_{ij}^2}\right) \tag{12}$$

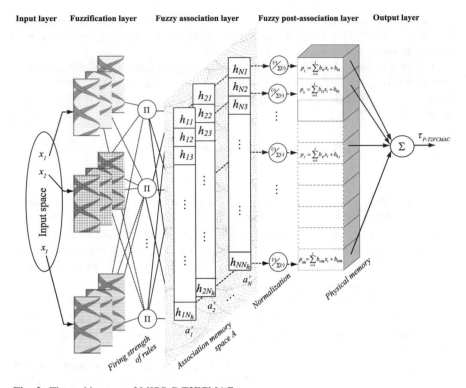

Fig. 2 The architecture of MISO P-T2FCMAC

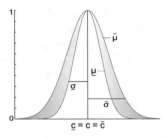

Fig. 3 Type-2 Gaussian fuzzy sets with uncertain standard deviation

where c_{ij} $(i = 1 \ldots I; j = 1 \ldots J)$ is the mean value of the jth fuzzy set for the ith input signal; $\overline{\sigma}_{ij}$ and $\underline{\sigma}_{ij}$ are deviations of the upper and lower membership functions of the jth fuzzy set for the ith input signal.

Layer 3. *Fuzzy association layer*: The layer is also called receptive-field space. Neurons in this layer are represented by the firing strength of the corresponding rule R_r calculated as the T-norm product of the degrees of fulfillment for both membership functions (upper and lower) respectively:

$$\overline{w}_r = \overline{\mu}_{\tilde{A}1}(x_1) * \overline{\mu}_{\tilde{A}2}(x_2) * \cdots * \overline{\mu}_{\tilde{A}I}(x_I)$$
$$\underline{w}_r = \underline{\mu}_{\tilde{A}1}(x_1) * \underline{\mu}_{\tilde{A}2}(x_2) * \cdots * \underline{\mu}_{\tilde{A}I}(x_I)$$
(13)

Moreover, memory element selection (association) vectors a_r^T ought to be determined in this layer, according the idea proposed in [6]. It was mentioned before that the activated cells formed by shifting units are called receptive fields or hyper-cubes [26]. Each location corresponds to a fuzzy association given by (10).

For the P-T2FCMAC network with two inputs, introduced in Fig. 2, type-2 fuzzy sets with three Gaussian membership functions (Fig. 3) are imposed on each input variable x_i. Accordingly, there are 9 fuzzy rules with 9 association vectors, a_r^T, ($r = 1, 2, \ldots 9$). The reduction of the association vectors' number is obvious here. In the CMAC network there are 16 vectors, while fuzzy reasoning determines 9 association vectors. The logical operation 'OR' is performed on all possible (in the same region) association vectors in CMAC.

Layer 4. *Fuzzy post-association layer*: It is also called weight memory space. The layer performs normalization of the firing strengths calculated in the previous layer [26]:

$$\tilde{\underline{w}}_r = \frac{\underline{w}_r}{\sum_{r=1}^{N} \underline{w}_r}, \quad \tilde{\overline{w}}_r = \frac{\overline{w}_r}{\sum_{r=1}^{N} \overline{w}_r}$$
(14)

Furthermore, the output weights of this layer correspond to the weighting column vector P of the P-T2FCMAC.

Layer 5. *Output layer*: The last layer of the P-T2FCMAC network performs type reduction and defuzzification operations. It evaluates the P-T2FCMAC output $\tau_{P-T2FCMAC}$ in accordance with the proposed in [4] type-2 fuzzy inference engine.

$$\tau_{P-T2FCMAC} = \eta \sum_{r=1}^{N} a_r^T P \tilde{\underline{w}}_r + (1 - \eta) \sum_{r=1}^{N} a_r^T P \tilde{\overline{w}}_r$$
(15)

where N is the number of the fuzzy rules; $\tilde{\underline{w}}_r$ and $\tilde{\overline{w}}_r$ are normalized values of the firing strengths (13) calculated according to (14).

The parameter η is introduced to address the case of non-uniformed uncertainties. It allows performing on-line adjustment of the influence of the lower or the upper membership function of IT2-FSs on the output determination procedure.

In order to show how the type-2 fuzzy reasoning is implemented into the basic CMAC architecture (Fig. 1), it is convenient to define the output of the proposed P-T2FCMAC in a matrix form (15):

$$\tau_{P-T2FCMAC} = \left(\eta \tilde{\underline{W}} A + (1 - \eta) \tilde{\overline{W}} A \right) P$$
(14)

where $\widetilde{\overline{W}} = [\widetilde{\overline{w}}_1 \ \widetilde{\overline{w}}_2 \ldots \widetilde{\overline{w}}_r \ldots \widetilde{\overline{w}}_N]$ and $\widetilde{\underline{W}} = [\widetilde{\underline{w}}_1 \ \widetilde{\underline{w}}_2 \ldots \widetilde{\underline{w}}_r \ldots \widetilde{\underline{w}}_N]$. The novelty of the proposed here fuzzy CMAC structure is related to the fuzzy association layer, where IT2-FSs are used to perform a nonlinear input-output mapping.

3.3 The Sliding Mode On-Line Learning Algorithm

An on-line algorithm based on the variable structure systems (VSSs) theory is proposed for the parameters tuning of the considered P-T2FCMAC network. It includes adaptation of three groups of network parameters: (i) the parameters of IT2 Gaussian functions, (ii) the parameters of the linear functions in the consequent parts of the fuzzy rules and (iii) the parameter η.

Let us define the learning error of the P-T2FCMAC network as the difference between the network's current output $\tau_{P-T2FCMAC}$ and its desired value τ^*:

$$\varepsilon(t) = \tau_{P-T2FCMAC}(t) - \tau^*(t) \tag{15}$$

It will be assumed that the rates of change of the desired output, $\dot{\tau}^*(t)$ and of the input signals, $\dot{x}_i(t)$ are bounded by a predefined positive constants $B_{\tau*\_dot}$ and B_{x_dot} (this limitation is valid for all real signal sources due to the physical restrictions).

$$\begin{aligned} |\dot{\tau}*(t)| &\leq B_{\tau*\_dot} & \forall t \\ |\dot{x}_i(t)| &\leq B_{x\_dot}, & (i = 1\ldots I) \ \forall t \end{aligned} \tag{16}$$

It will be assumed also that, the time-varying coefficients b_{ij} in the consequents part of fuzzy *if-then* rules of the P-T2CMAC network are bounded, i.e.,

$$|b_{ij}(t)| \leq B_b \quad (i = 0, 1\ldots I, \ j = 1\ldots N_h) \quad \forall t \tag{17}$$

Based on the principles of the sliding mode control theory [25] the zero value of the learning error can be defined as a time-varying sliding surface, i.e.

$$s(\varepsilon(t)) = \varepsilon(t) = \tau_{P-T2FCMAC}(t) - \tau^*(t) = 0 \tag{18}$$

which is the condition that guarantees that when the system is in a sliding mode on the sliding surface s the P-T2FCMAC neural network output $\tau_{P-T2FCMAC}(t)$ will coincide with the desired output signal $\tau^*(t)$ for all time $t > t_h$ where t_h is the hitting time when $\varepsilon(t) = 0$.

Definition 1 A sliding motion will have place on a sliding manifold $s(\varepsilon(t)) = \varepsilon(t) = 0$, after time t_h if the condition $s(t)\dot{s}(t) = \varepsilon(t)\dot{\varepsilon}(t) < 0$ is true for all t in some nontrivial semi open subinterval of time of the form $[t, t_h) \subset (-\infty, t_h)$.

The algorithm for the adaptation of the parameters $\bar{c}_{ij}, \underline{c}_{ij}, \bar{\sigma}_{ij}, \underline{\sigma}_{ij}, b_{ij}, b_{0j}, \eta$ should be derived in such a way that the sliding mode condition of the above Definition 1 will be enforced.

Theorem 1 *If the learning algorithm for the parameters of the upper $\bar{\mu}(x_i)$ and the lower $\underline{\mu}(x_i)$ membership functions with a Gaussian distribution is chosen respectively as:*

$$\underline{\dot{c}}_{ij} = \bar{\dot{c}}_{ij} = \dot{x}_i$$
$$\underline{\dot{\sigma}}_{ij} = -\alpha\,sign(\varepsilon)\frac{\underline{\sigma}_{ij}^3}{(x_i-c_{ij})^2} \quad \bar{\dot{\sigma}}_{ij} = -\alpha\,sign(\varepsilon)\frac{\bar{\sigma}_{ij}^3}{(x_i-c_{ij})^2} \tag{19}$$

and the adaptation of the coefficients in the consequents part of fuzzy rules is chosen as:

$$\dot{b}_{ij} = -\alpha\,sign(\varepsilon)/x_i; \quad \dot{b}_{0j} = -\alpha\,sign(\varepsilon)\frac{1}{A\left(\eta\underline{\tilde{W}} + (1-\eta)\overline{\widetilde{W}}\right)} \tag{20}$$

and the weight coefficient η is updated as follows:

$$\dot{\eta} = -\alpha\,sign(\varepsilon)\frac{1}{AP\left(\underline{\tilde{W}} - \overline{\widetilde{W}}\right)^T} \tag{21}$$

where $\underline{\tilde{W}} = [\underline{\tilde{w}}_1 \ldots \underline{\tilde{w}}_N]; \overline{\widetilde{W}} = [\overline{\widetilde{w}}_1 \ldots \overline{\widetilde{w}}_N]; \alpha$ is a sufficiently large positive constant satisfying the inequality:

$$\alpha > (IB_bB_{x\_dot} + B_{\tau^*\_dot})/(I+2) \tag{22}$$

then, given an arbitrary initial condition $\varepsilon(0)$, the learning error $\varepsilon(t)$ will converge to zero during a finite time t_h.

Proof From (7), (8) and (9) it is possible to obtain the time derivatives:

$$\underline{\dot{\tilde{w}}}_r = -\underline{\tilde{w}}_r\underline{K}_r + \underline{\tilde{w}}_r\sum_{r=1}^{N}\underline{\tilde{w}}_r\underline{K}_r; \quad \overline{\dot{\widetilde{w}}}_r = -\overline{\widetilde{w}}_r\overline{K}_r + \overline{\widetilde{w}}_r\sum_{r=1}^{N}\overline{\widetilde{w}}_r\overline{K}_r \tag{23}$$

where the following substitutions are used

$$\underline{K}_r = \sum_{i=1}^{I}\left[\left(\frac{x_i-c_{ij}}{\underline{\sigma}_{ij}}\right)\left(\frac{x_i-c_{ij}}{\underline{\sigma}_{ij}}\right)'\right]; \quad \overline{K}_r = \sum_{i=1}^{I}\left[\left(\frac{x_i-c_{ij}}{\bar{\sigma}_{ij}}\right)\left(\frac{x_i-c_{ij}}{\bar{\sigma}_{ij}}\right)'\right] \tag{24}$$

It is also obvious that by applying the proposed adaptation laws (19) the value of $\underline{K}_r$ and $\overline{K}_r$ can be calculated as follows:

$$\overline{K}_r = \underline{K}_r = I\alpha sign(\varepsilon) \tag{25}$$

Consider $V = \frac{1}{2}\varepsilon^2$ as Lyapunov function candidate.

$$\dot{V} = \varepsilon\dot{\varepsilon} = \varepsilon(\dot{\tau}_{P-T2FCMAC} - \dot{\tau}^*) \tag{26}$$

By differentiating (15) it is possible to obtain:

$$
\begin{aligned}
\dot{\tau}_{P-T2CFCMAC} &= \dot{\eta}\sum_{r=1}^{N} a_r^T b\underline{\tilde{w}}_r + \eta\sum_{r=1}^{N}\left((\dot{a}_r^T P + a_r^T \dot{P})\underline{\tilde{w}}_r + a_r^T P\underline{\dot{\tilde{w}}}_r\right) \\
&\quad - \dot{\eta}\sum_{r=1}^{N} f_r\overline{\tilde{w}}_r + (1-\eta)\sum_{r=1}^{N}\left((\dot{a}_r^T P + a_r^T \dot{P})\overline{\tilde{w}}_r + a_r^T P\dot{\overline{\tilde{w}}}_r\right) \\
&= \dot{\eta}\sum_{r=1}^{N} f_r\underline{\tilde{w}}_r + \eta\sum_{r=1}^{N}\left(\dot{f}_r\underline{\tilde{w}}_r + f_r\underline{\dot{\tilde{w}}}_r\right) \\
&\quad - \dot{\eta}\sum_{r=1}^{N} f_r\overline{\tilde{w}}_r + (1-\eta)\sum_{r=1}^{N}\left(\dot{f}_r\overline{\tilde{w}}_r + f_r\dot{\overline{\tilde{w}}}_r\right)
\end{aligned} \tag{27}
$$

Note that the sums of the normalized activations are units:

$$\sum_{r=1}^{N}\underline{\tilde{w}}_r = 1; \quad \sum_{r=1}^{N}\overline{\tilde{w}}_r = 1 \tag{28}$$

Hence, the (27) can be rewritten as:

$$
\begin{aligned}
\dot{\tau}_{P-T2FCMAC} &= -\dot{\eta}\sum_{r=1}^{N} a_r^T P\left(\underline{\tilde{w}}_r - \overline{\tilde{w}}_r\right) \\
&\quad + \sum_{r=1}^{N}\left(\dot{a}_r^T P + a_r^T \dot{P}\right)\left(\eta\underline{\tilde{w}}_r + (1-\eta)\overline{\tilde{w}}_r\right) = -\dot{\eta}\sum_{r=1}^{N} a_r^T P\left(\underline{\tilde{w}}_r - \overline{\tilde{w}}_r\right) \\
&\quad + \sum_{r=1}^{N}\left(\sum_{j=1}^{Nh} h_{rj}\dot{p}_j\left(\eta\underline{\tilde{w}}_r + (1-\eta)\overline{\tilde{w}}_r\right)\right) = -\dot{\eta}\sum_{r=1}^{N} a_r^T P\left(\underline{\tilde{w}}_r - \overline{\tilde{w}}_r\right) \\
&\quad + \sum_{r=1}^{N}\left(\sum_{j=1}^{Nh} h_{rj}\left(\sum_{i=1}^{I}\left(\dot{b}_{ij}x_i + b_{ij}\dot{x}_i\right) + \dot{b}_{0j}\right)\left(\eta\underline{\tilde{w}}_r + (1-\eta)\overline{\tilde{w}}_r\right)\right)
\end{aligned} \tag{29}
$$

By applying (19), (20) and (21), the following expression is obtained:

$$\dot{\tau}_{P-T2FCMAC} = -\alpha\, sign(\varepsilon)\frac{1}{AP\left(\underline{\widetilde{W}} - \widetilde{\widetilde{W}}\right)^T}\sum_{r=1}^{N} a_r^T P(\underline{\widetilde{w}}_r - \widetilde{\widetilde{w}}_r)$$

$$- \alpha\, sign(\varepsilon)\frac{1}{A\left(\eta\underline{\widetilde{W}} + (1-\eta)\widetilde{\widetilde{W}}\right)}\sum_{r=1}^{N} a_r^T\left(\eta\underline{\widetilde{w}}_r + (1-\eta)\widetilde{\widetilde{w}}_r\right)$$

$$+ \sum_{r=1}^{N}\left(\sum_{j=1}^{Nh} h_{rj}\left(\sum_{i=1}^{I}(\dot{b}_{ij}x_i + b_{ij}\dot{x}_i)\right)\left(\eta\underline{\widetilde{w}}_r + (1-\eta)\widetilde{\widetilde{w}}_r\right)\right) \tag{30}$$

By substituting (30) into (26) the derivative of the Lyapunov function candidate can be calculated as follows.

$$\dot{V} = \varepsilon\dot{\varepsilon} = \varepsilon(\dot{\tau}_{P-T2FCMAC} - \dot{\tau}^*)$$

$$= \varepsilon\left[-2\alpha\, sign(\varepsilon) + \sum_{r=1}^{N}\left(\sum_{j=1}^{Nh} h_{rj}\left(\sum_{i=1}^{I}\dot{b}_{ij}x_i + b_{ij}\dot{x}_i\right)\left(\eta\underline{\widetilde{w}}_r + (1-\eta)\widetilde{\widetilde{w}}_r\right)\right) - \dot{\tau}^*\right]$$

$$\dot{V} = \varepsilon\left[-2\alpha\, sign(\varepsilon) + \sum_{r=1}^{N}\left(\sum_{j=1}^{Nh} h_{rj}(-I\alpha sign(\varepsilon))\right.\right.$$

$$\left.\left. + \sum_{i=1}^{I} b_{ij}\dot{x}_i\right)\left(\eta\underline{\widetilde{w}}_r + (1-\eta)\widetilde{\widetilde{w}}_r\right)\right)\right]$$

$$\dot{V} = -|\varepsilon|\left[2\alpha + I\alpha\left(\eta\sum_{r=1}^{N} a_r^T\underline{\widetilde{w}}_r + (1-\eta)\sum_{r=1}^{N} a_r^T\widetilde{\widetilde{w}}_r\right)\right]$$

$$+ \varepsilon\left[\eta\sum_{r=1}^{N}\left(\underline{\widetilde{w}}_r\sum_{j=1}^{Nh} h_{rj}\sum_{i=1}^{I} b_{ij}\dot{x}_i\right) + (1-\eta)\sum_{r=1}^{N}\left(\widetilde{\widetilde{w}}_r\sum_{j=1}^{Nh} h_{rj}\sum_{i=1}^{I} b_{ij}\dot{x}_i\right) - \dot{\tau}^*\right] \tag{31}$$

$$\dot{V} < -|\varepsilon|[\alpha(I+2)] + |\varepsilon|\left[\begin{array}{c}\eta\sum_{r=1}^{N}\left(\underline{\widetilde{w}}_r\sum_{j=1}^{Nh} h_{rj}\sum_{i=1}^{I} B_b B_{x\_dot}\right)\\[2mm] + (1-\eta)\sum_{r=1}^{N}\left(\widetilde{\widetilde{w}}_r\sum_{j=1}^{Nh} h_{rj}\sum_{i=1}^{I} B_b B_{x\_dot}\right) + B_{\tau^*\_dot}\end{array}\right]$$

$$= -|\varepsilon|\alpha(I+2) + |\varepsilon|\left[\begin{array}{c}\eta I B_b B_{x\_dot}\sum_{r=1}^{N}\underline{\widetilde{w}}_r a_r^T\\[2mm] + (1-\eta)I B_b B_{x\_dot}\sum_{r=1}^{N}\widetilde{\widetilde{w}}_r a_r^T + B_{\tau^*\_dot}\end{array}\right]$$

$$= -|\varepsilon|\alpha(I+2) + |\varepsilon|(I B_b B_{x\_dot} + B_{\tau^*\_dot})$$

$$= |\varepsilon|(I B_b B_{x\_dot} + B_{\tau^*\_dot} - \alpha(I+2)) < 0 \tag{32}$$

The last inequality is true if (22) is satisfied. The inequality (32) means that the controlled trajectories of the learning error $\varepsilon(t)$ converge to zero in a stable manner.

A standard approach to avoid the chattering phenomenon (which is a well-known problem associated with SMC) is to smooth the discontinuity introduced by the signum function in (19)–(21) by using the following substitution

$$sign(\varepsilon(t)) \approx \frac{\varepsilon(t)}{|\varepsilon(t)| + \delta} \tag{33}$$

where δ is a small positive scalar.

4 The Trajectory Tracking Control Scheme

By applying the discussed P-T2FCMAC network as a friction and disturbances compensator its output, $\tau_{P-T2FCMAC}$ will be used to estimate the value of the sum of $F(\dot{q})$ and τ_d terms in (5) in order to calculate the appropriate control input τ. The input signals of the network are the position error and its time derivative $x = [e\ \dot{e}]$. Thus (5) becomes

$$\tau = M(q)(\ddot{q}_d - u) + V_m(q, \dot{q})\dot{q} + G(q) + \tau_{P-T2FCMAC} \tag{34}$$

The perfect compensation will be obtained if $\tau_{P-T2FCMAC}$ represents the exact model of the friction and disturbances $\tau_{P-T2FCMAC} = F(\dot{q}) + \tau_d$. In order to apply the sliding mode adaptation algorithm (19)–(21) the error between the current output of the compensator $\tau_{P-T2FCMAC}$ and its desired value τ^* is needed.

$$\varepsilon(t) = \tau_{P-T2FCMAC}(t) - \tau^*(t) \tag{35}$$

Since the friction and disturbances are unknown so the desired value τ^* is also unknown. Thus we will apply the filtered error approach for manipulator trajectory control [17]. Let define

$$r = \dot{e} + \Lambda e \tag{36}$$

where $\Lambda > 0$ is a design parameter matrix and r is the filtered error. The dynamical system (36) is stable so $e(t)$ is bounded as long as the controller guarantees that $r(t)$ is bounded [19]. The manipulator dynamics may be written in terms of the filtered tracking error as

$$M\dot{r} = -V_m r - \tau + f \tag{37}$$

where f is a nonlinear function

$$f = M(\ddot{q}_d + \Lambda\dot{e}) + V_m(\dot{q}_d + \Lambda e) + G + F + \tau_d \tag{38}$$

The adopted estimation of f is

$$\hat{f} = M(\ddot{q}_d + \Lambda\dot{e}) + V_m(\dot{q}_d + \Lambda e) + G + \tau_{P-T2FCMAC} \tag{39}$$

If we define a control input torque as

$$\tau = \hat{f} + K_v r \tag{40}$$

with gain matrix $K_v = K_v^T > 0$ then (37) becomes

$$M\dot{r} = -(K_v + V_m)r + F + \tau_d - \tau_{P-T2FCMAC} \tag{41}$$

If the desired output of the adaptive compensator τ^* is chosen as

$$\tau^*(t) = r = e + \Lambda\dot{e} \tag{42}$$

and applying the sliding mode learning algorithm (19)–(21) it follows that $\tau_{P-T2FCMAC}(t) \rightarrow r(t)$. Then (41) can be written as

$$\dot{r} + M^{-1}(K_v + V_m + 1)r = M^{-1}(F + \tau_d) \tag{43}$$

The dynamical system (43) is stable as long as the friction term and disturbances are bounded. This means filtered error $r(t)$ remains bounded and thus the tracking error $e(t)$ and its time-derivative $\dot{e}(t)$ are also bounded.

The proposed trajectory tracking control scheme with the adaptive P-T2FCMAC friction and disturbance compensator is shown in Fig. 4.

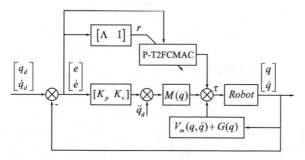

Fig. 4 Architecture of the trajectory tracking control scheme with the proposed adaptive P-T2FCMAC friction and disturbance compensator

5   Simulation Results

The performance of the proposed adaptive P-T2FCMAC neural compensator is evaluated when applied on a two-link RR planar manipulator shown in Fig. 5. It has been modeled as two rigid links of length 1 m each with masses $m_1 = 1$ kg and $m_2 = 1$ kg concentrated at the distal ends of the links. The following nonlinear viscous and dynamic friction terms of $F(\dot{q})$ and unknown disturbances τ_d have been included in the manipulator dynamics.

$$F(\dot{q}) = F_v \dot{q} + F_d \, sign(\dot{q})$$
$$\tau_d = T_d \sin(15t) \tag{44}$$

where $F_v = 1$; $F_d = 0.5$; $T_d = 0.8$.

The desired trajectory to be followed is given by

$$q_{d1}(t) = \sin(2\pi t/T) + 1; \quad q_{d2} = \cos(2\pi t/T) + 1 \tag{45}$$

where $T = 2$ s. The matrix K_v is taken as $K_v = [100 \ 0; 0 \ 20]$. The design parameter Λ is taken as diagonal matrix with entries equal to 3. In the simulation experiments the performance of the trajectory tracking control scheme with the proposed adaptive P-T2FCMAC friction and disturbance compensator is compared to the cases when adaptive type-1 and type-2 fuzzy neural networks (presented in [23, 24]) are used as compensators, and also without an intelligent compensator. Figures 6 and 7 show the tracking performance and the tracking error of link 1 and link 2 respectively.

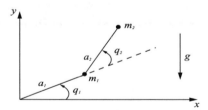

Fig. 5 Two-link planar manipulator

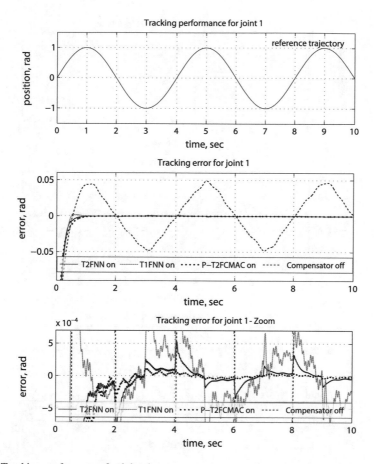

Fig. 6 Tracking performance for joint 1

It can be seen from the simulations that the robot manipulator tracking performance is much better when the proposed adaptive P-T2FCMAC friction and disturbance compensator is turned on. The root-mean-squared-error values obtained for the three investigated intelligent friction and disturbance compensators are presented in Table 1.

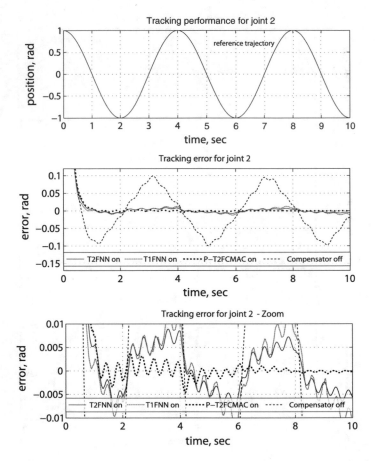

Fig. 7 Tracking performance for joint 2

Table 1 RMSE values of		RMSE/joint1	RMSE/joint2
tracking performance	T1FNN	0.0170	0.1154
	T2FNN	0.0166	0.1144
	P-T2FCMAC	0.0159	0.1123

6 Conclusions

An on-line tuned parametric type-2 fuzzy CMAC network is proposed for compensation of friction and disturbances during the trajectory tracking control of robot manipulators. The parameter adaptation applies a newly developed stable sliding mode-based on-line learning algorithm. The control approach guarantees the stability of closed loop system and the simulation results demonstrate an improved performance in the presence of disturbances and unmodeled friction.

References

1. S. Ahmed, K. Shiev, A.V. Topalov, N. Shakev, O. Kaynak, Sliding mode online learning algorithm for type-2 fuzzy CMAC networks, in *Proceedings of the 9th Asia Control Conference*, Istanbul, Turkey (to be published) (2013)
2. J.S. Albus, A new approach to manipulator control: the cerebellar model articulation controller (CMAC). ASME J. Dyn. Syst. Measur. Control, 220–227 (1975)
3. J.S. Albus, Data storage in the cerebellar model articulation controller (CMAC). ASME J. Dyn. Systems, Measur. Control, 228–233 (1975)
4. M. Biglarbegian, W. Melek, J.M. Mendel, On the stability of interval type-2 TSK fuzzy logic control systems. Syst. Man Cybern. Part B Cybern. IEEE Trans. **40**(3), 798–818 (2010)
5. C.S. Chen, Sliding-mode-based fuzzy CMAC controller design for a class of uncertain nonlinear systems, in *Proceedings of the 2009 IEEE International Conference on Systems, Man, and Cybernitics*, San Antonio, USA, pp. 3030–3034 (2009)
6. J.Y. Chen, P.S. Tsai, C.C. Wong, Adaptive design of a fuzzy cerebellar model arith-metic controller neural network. IEE Proc. Control Theory Appl. **152**(6), 133–137 (2005)
7. K.H. Cheng, CMAC-based neuro-fuzzy approach for complex system modeling. Neurocomputing **72**, 1763–1774 (2009)
8. M.A. Costa, A.P. Braga, B.R. Menezes, Improving generalization of MLPs with sliding mode control and the Levenberg-Marquardt algorithm. Neurocomputing **70**(7–9), 1342–1347 (2007)
9. E.G. Gilbert, I.J. Ha, An approach to nonlinear feedback control with applications to robotics. IEEE Trans. Syst. Man Cybern. **14**(6), 879–884 (1984)
10. H.A. Hagras, A hierarchical type-2 fuzzy logic control architecture for autonomous mobile robots. IEEE Trans. Fuzzy Syst. **12**(4), 524–539 (2004). doi:10.1109/TFUZZ.2004.832538
11. V.M. Hung, U.J. Na, Adaptive neural fuzzy control for robot manipulator friction and disturbance compensator, in *International Conference on Control, Automation and Systems 2008*, COEX, Seoul, Korea, pp. 2569–2574, 14–17 Oct 2008
12. L.R. Hunt, R. Su, G. Meyer, Global transformations of nonlinear systems. IEEE Trans. Autom. Control **28**(1), 24–31 (1983)
13. K.S. Hwang, C.S. Lin, Smooth trajectory tracking of three-link robot: a self-organizing CMAC approach. IEEE Trans. Syst. Man Cybern. B **28**, 680–692 (1998)
14. M.A. Khanesar, E. Kayacan, M. Teshnehlab, O. Kaynak, Analysis of the noise reduction property of type-2 fuzzy logic systems using a novel type-2 membership function. IEEE Trans. Syst. Man Cybern. **41**(5), 1395–1405 (2011)
15. S.H. Lane, D.A. Handelman, J.J. Gelfand, Theory and development of higher-order CMAC neural networks. IEEE Contr. Syst. Mag. **12**, 23–30 (1992)
16. C.H. Lee, C.C. Teng, Identification and control of dynamic systems using recurrent fuzzy neural network. IEEE Trans. Fuzzy Syst. **8**, 349–366 (2000)
17. F.L. Lewis, J. Campus, R. Semic, *Neuro-Fuzzy Control of Industrial Systems with Actuator Nonlinearities* (SIAM, Philadelphia, 2002)
18. F.L. Lewis, D.M. Dawson, C.T. Abdallah, *Robot Manipulator Control. Theory and Practice*, 2nd edn. (Marcel Dekker, Inc., New York, Basel, 2004)
19. F.L. Lewis, S. Jagannathan, A. Yesildirek, *Neural Network Control of Robot Manipulators and Nonlinear Systems* (Taylor & Francis, London, Philadelphia, 1999)
20. Z. Man, H.R. Wu, S. Liu, X. Yu, A new adaptive back-propagation algorithm based on Lyapunov stability theory for neural networks. IEEE Trans. Neural Netw. **17**(6), 1580–1591 (2006)
21. J.M. Mendel, *Uncertain Rule-Based Fuzzy Logic Systems: Introduction and New Directions* (Prentice-Hall, Englewood Cliffs, 2001)
22. J.M. Mendel, R. John, Type-2 fuzzy sets made simple. IEEE Trans. Fuzzy Syst. **10**(2), 117–127 (2002)

23. K. Shiev, N. Shakev, A.V. Topalov, S. Ahmed, Trajectory control of manipulators using type-2 fuzzy neural friction and disturbance compensator, in *Intelligent Systems (IS), 2012 6th IEEE International Conference*, pp. 324–329 (2012). doi:10.1109/IS.2012.6335155
24. A.V. Topalov, A new stable on-line learning for takagi-sugeno-kang type neuro-fuzzy networks. Comptes rendus de l'Academie bulgare des Sciences 64(10), 1489–1498 (2011)
25. V.I. Utkin, *Sliding Modes in Control and Optimization* (Springer, Berlin, 1992)
26. T.F. Wu, P.S. Tsai, F.R. Chang, L.S. Wang, Adaptive fuzzy CMAC control for a class of nonlinear systems with smooth compensation. IEE Proc. Control Theory Appl. **153**(6), 647–657 (2006)
27. W. Yu, X. Li, Fuzzy identification using fuzzy neural networks with stable learning algorithms. IEEE Trans. Fuzzy Syst. **12**(3), 411–420 (2004)

On Heading Change Measurement: Improvements for Any-Angle Path-Planning

Pablo Muñoz and María D. R-Moreno

Abstract Finding the most efficient and safe path between locations is a ubiquitous problem that occurs in smart phone GPS applications, mobile robotics and even video games. Mobile robots in particular must often operate in any type of terrain. The problem of finding the shortest path on a discretized, continuous terrain has been widely studied, and many applications have been fielded, including planetary exploration missions (i.e. the MER rovers). In this chapter we review some of the most well known path-planning algorithms and we propose a new parameter that can help us to compare them under a different measure: *the heading changes* and to perform some improvements in any-angle path-planning algorithms. First, we define a heuristic function to guide the process towards the objective, improving the computational cost of the search. Results show that algorithms using this heuristic get better runtime and memory usage than the former ones, with a slightly degradation of other parameters such as path length. And second, we modify an any-angle path-planning algorithm to consider heading changes during the search in order to minimize them. Experiments show that this algorithm obtains smoother paths than the other algorithms tested.

1 Introduction

Path-planning is a subtask of motion planning and a widely research problem in mobile robotics and video games. It is focused on finding an obstacle-free path between an initial position and a goal, trying as far as possible that this path is optimal. The path-planning problem representation as a search tree over discrete environments with blocked or unblocked squares has been extensively discussed.

P. Muñoz (✉) · M.D. R-Moreno
Departamento de Automática, Universidad de Alcalá, Alcalá de Henares, Madrid, Spain
e-mail: pmunoz@aut.uah.es

M.D. R-Moreno
e-mail: mdolores@aut.uah.es

© Springer International Publishing Switzerland 2016
M. Hadjiski et al. (eds.), *Novel Applications of Intelligent Systems*,
Studies in Computational Intelligence 586, DOI 10.1007/978-3-319-14194-7_5

It is usual to assume (in this chapter we do it) that the terrain is *fully-observable*, so we know about all the obstacles in the terrain, and *static*, that is, no new obstacles are added during the path extraction.

Algorithms for path-planning are usually high consumption processes in terms of CPU load and memory required. Both, processor and memory are a bottleneck for search algorithms that may grow exponentially with the problem complexity [1]. This implies difficulties for the typical applications of these algorithms:

- In video games the number of units to compute paths could be enough to "freezes" the computer when they require to obtain a path as Higghins described [2]. Games are interactive systems, and when a player wants to make a move, he expects that the movement starts immediately. A delay in the execution response may result in a bad game experience and thus, bad sales.
- For robotics systems there is generally an integrated computer with low performance, where some time can be dedicated on searching a path, but the remaining time should be used on other important questions such as transmitting data, diagnosis, etc. Also, the robot must deal, generally, with an uncertain and dynamic environment. So reactivity is desirable in certain situations when the integrity of the robot could depend on how quickly can take a decision.

We can perform the path extraction via uninformed search, using algorithms such as Dijkstra [3], which only considers the cost of reaching a position as the evaluation function for the search. So, a bigger area of the search space must be explored to reach a path to the goal point that results in an expensive process. Informed search, instead, can take advantages of domain dependent heuristics, which combined with the cost value, give us a better evaluation function that guides the search towards the objective without the necessity of expanding all the possible positions.

In this chapter we take into consideration the measurement of heading changes during the search process to guide the search towards the objective in a greedy way. To do this, we consider that the shortest path between two points is the straight line if there are no obstacles. So, points far from this line are not desirable and thus we do not want to expand them: they probably lead us further away from the goal. Using this as part of the heuristic value allow us to minimize the number of expanded nodes, and thus, the processing time and the memory required during the search, but having a negative impact in the path length and heading changes.

On the other hand, if we consider the heading changes during the search as part of the cost function, we obtain a search algorithm that optimizes the combination of path length plus heading changes of the route, not only the path length as general path-planning algorithms do. An algorithm that provides smoother paths could be very desirable for some kind of robots with limitations on turning.

The chapter is structured as follows: next section presents a briefly introduction about classical path-planning, the notation employed and, in more detailed, the explanation of A* search algorithm applied to the path-planning problem, and two algorithms directly inherited from it (A*Post Processed and Theta*). Section 3 presents the definition and formulation for angle processing required in the following sections. Then, we apply these definitions to the search process as part of the heuristic function in Sect. 4 and as part of the cost function in Sect. 5. Next, we present the experimental result for different type and size maps. And finally, some conclusions are outlined.

2 Path-Planning Review

The classical algorithm to solve the path-planning problem is the A* search algorithm [4], which allows one to quickly find routes over a grid at the expense of an artificial restriction of heading changes of $\pi/4$. A* is simple to implement, is very efficient, and has lots of scope for optimization [5]. In this direction, there has been many improvements such as A* Post Smoothed (A*PS) [6] that smooths the path obtained by A* at a later stage; Field D* [7] a sophisticated modification that works on partially known and non-uniform costs maps using a replanning scheme in order to fast generation of new paths when an unknown obstacle is detected; or an approach that uses framed cells approach (a subdivision of the grid) for continuous-field path-planning with path-dependent state variables [8].

More recently has appeared the Theta* algorithm [9], which removes the restriction on heading changes that generates A* and gets better path lengths with any-angle headings. The main difference between A* and Theta* is that the former only allows that the parent of a node is its predecessor, while in the last, the parent of a node can be any node (if there is line of sight). This property allows Theta* to find shorter paths with fewer turns compared to A* [10]. However, this improvement implies a higher computational cost due to additional operations to be performed in the expansion nodes process required to check the line of sight between no adjacent nodes. In order to reduce the time spent by the line of sight calculation Choi et al. [11] include pruning rules that can reduce the runtime of Theta* by up to a factor close to 2 without a significant increase in the path length.

Finally, when comparing path-planning algorithms, it is common to contrast four parameters: length of the solution (path length), runtime, expanded nodes and number of heading changes. Usually in the literature the goodness of an algorithm is focused on the first two parameters. Daniel et al. [10] perform an extensive comparison between Theta*, A*, Field D* and A*PS. As a result we can see that Theta* is the one that usually gets the shortest routes.

In this chapter, we focus on A* and Theta*, that are actually the most well known and used path-planning algorithms over grids. Next sections present the nomenclature employed during the rest of the article and present both algorithms (plus A*PS) in more detail.

2.1   Grid Definition and Notation

The most common terrain discretization in path-planning is a regular grid with
blocked and unblocked square cells [12]. For this kind of grids we can find two
variants: (i) the center-node (Fig. 1 left) in which the mobile element is in the center
of the square; and (ii) corner-node (Fig. 1 right), where nodes are the vertex of the
square. For both cases, a valid path is that starting from the initial node reaches the
goal node without crossing a blocked cell. In our experiments we have employed
the corner-node approximation, but results presented can be applied to the
center-node representation.

A node is represented as a lowercase letter, assuming p a random node and, s and
g the start and goal nodes respectively. Each node is defined by its coordinate pair
(x, y), being X_p and y_p for the p node. A solution has the form $(p_1, p_2, \ldots, p_{n-1}, p_n)$
with initial node $p_1 = s$ and goal $p_n = g$. As well, we have defined three functions
related to nodes: (i) function $succ\ (p)$ that returns a subset with the visible neighbors
of p; (ii) function $parent\ (p)$ that indicates who is the parent node of p; and (iii) $dist$
(p, t) that represents the straight line distance between nodes p and t, calculated
through the Eq. 1.

$$\text{dist}(p, t) = \sqrt{(x_t - x_p)^2 + (y_t - y_p)^2} \qquad (1)$$

2.2   A* Algorithm for Path-Planning

Some path-planning algorithms are a variation of the A* search algorithm [4]. It is
simple to implement, is very efficient, optimal when it is applied to visibility graphs
and has lots of scope for optimization [5]. But it has an important limitation: is
typical to use 8 neighbors nodes, so this restricts the path headings to multiples of $\pi/4$, causing that A* generates a sub-optimal path with artificial zig-zag patterns. It is
possible to use more adjacent nodes to relax this constraint, but the complexity
grows exponentially with each increment.

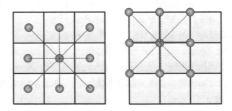

Fig. 1 Possible node representations in *grids*. *Left* center-node; *Right* corner-node

For each node, A* maintains four important values:

- $G(t)$: the cumulative cost to reach the node t from the initial node, that is, the length of the shortest path from the start node to p node.
- $H(t)$: the heuristic value, i.e., the estimated distance to the goal node. A* uses the Octile heuristic, which represents the minimal cost to reach a node using an eight-grid connected, calculated through the Eq. 2.

$$\text{Octile}(p, t) = \sqrt{2} \cdot dsm + (dlg - dsm)$$
$$\text{with } dsm = min(|x_t - x_p|, |y_t - y_p|) \qquad (2)$$
$$\text{and } dlg = max(|x_t - x_p|, |y_t - y_p|)$$

- $F(t)$: the node evaluation function that encapsulates the G and H values, is expressed by Eq. 3.

$$F(t) = G(t) + H(t) \qquad (3)$$

- *parent* (t): the reference to the parent node. The parent of a node in A* must satisfy the following constraint due to the search process: $p = parent(t) \Rightarrow t \in succ(p)$.

A* has two lists to manage the search: (i) the *open* list, a priority queue of nodes ordered by their F values, and (ii) the *closed* list, that contains the nodes that have been already expanded. The search process, shown in Algorithm 1, will expand the most promising nodes in the order established in the open list, i.e. it first expands the nodes with lower values for F. In the case that the expanded node is the goal, the algorithm will return the path by traversing the parents pointers backwards from the goal to the start node. If instead the open list is empty, it means that it is impossible to reach the goal node from the initial node and the algorithm will return a failure.

When we are dealing with the successor list, *succ* (p) being p the current node, we need to update the data of non expanded nodes. This is done by the *Update Vertex* function at line 18. When the search reaches a node for the first time, or the cost to reach that node from the current position is less than the previously obtained, the G and H values are updated properly and the parent of that node is set to the current node. Then, we update the *open* list and the search process continues. The pseudo-code of this function is presented in the Algorithm 2, taking only into consideration the lines 12–19.

2.3 A* Post-processing: Improving A* Paths

There are some variations of A* to convert it into an any-angle algorithm [13, 14]. In this chapter we use A* Post Smoothed (abbreviated A*PS) algorithm described in [6]. It runs A* and then smooths the resulting path in a post-processing step. Therefore, the resultant path may be shorter than the original, but it increases the runtime. If A* finds a path $(p_1, p_2, ..., p_n)$, the smooth process checks the line of sight between the first node and the successor node of its successors. The meaning of line of sight between two nodes is that, if there are no obstacles in the straight line which connects these two nodes, then, there is a line of sight, otherwise, there is not. For example, taking the initial node $s = p_1$ as the current node, it checks if there is a line of sight between p_1 and p_3. If it is true, the parent of p_3 is now p_1 and thus p_2 is eliminated from the path. The algorithm then takes the next node in the path and checks the line of sight with the current node. If there is not visibility between these two nodes, the last node becomes the current node and the line of sight check continues. The process is repeated until it reaches the goal. So, the resultant path has the same or less nodes than the original one, that is, $n \geq j$, being n the number of nodes in the original path and j the nodes in the post smoothed path.

Algorithm 1 A* search

```
 1  G(s) ← 0
 2  parent(s) ← s
 3  open ← ∅
 4  open.insert(s, G(s), H(s))
 5  closed ← ∅
 6  while open ≠ ∅ do
 7      p ← open.pop()
 8      if p = g then
 9          return path
10      end if
11      closed.insert(p)
12      for t ∈ succ(p) do
13          if t ∉ closed then
14              if t ∉ open then
15                  G(t) ← ∞
16                  parent(t) ← null
17              end if
18              UpdateVertex(p, t)
19          end if
20      end for
21  end while
22  return fail
```

2.4 Theta* Algorithm: Any-Angle Path-Planning

Theta* [10, 9] is a variation of A* for any-angle path-planning on grids. There are two variants for Theta*: Angle-Propagation Theta* [9] and Basic Theta* [10]. We assume that talking about Theta* refers to the last one. Theta* is identical to

A* except the *UpdateVertex* function, so the pseudo-code for A* shown in Algorithm 1 applies also to Theta*.

Theta* works like A*PS: both try to connect no adjacent nodes that have line of sight, erasing intermediate nodes and thus, possible zig-zag patterns generated by A*. But there is an important difference between Theta* and A*PS: first one does not need a post-processing step, it does the line of sight check during the expansion of nodes. When Theta* expands a node, p, it checks the line of sight between the parent of the node and its successors. If there is line of sight between a successor of p and its parent, then the parent of the successor is *parent(p)*, not p like A*. When there is an obstacle blocking the line of sight, then Theta* works like A*. For this reason, the parent of a node can be any node, and the path obtained is no restricted to $\pi/4$ headings. The *UpdateVertex* function pseudocode for Theta* is shown in Algorithm 2. In order to obtain a better result, the heuristic that it employs is the Euclidean distance that represents the minimum cost to reach a node in a free obstacle area and without restrictions of heading changes. The formulation to compute the Euclidean distance is seen in Eq. 1. As a consequence of the expansion process, Theta* only has heading changes at the corners of the blocked cells.

Algorithm 2 Update vertex function for Basic Theta*

```
 1  UpdateVertex(p, t)
 2  if LineOfSight(parent(p),t) then
 3      if G(parent(p)) + dist(parent(p),t) < G(t) then
 4          G(t) ← G(parent(p)) + dist(parent(p),t)
 5          parent(t) ← parent(p)
 6          if t ∈ open then
 7              open.remove(t)
 8          end if
 9          open.insert(t, G(t), H(t))
10      end if
11  else
12      if G(p) + dist(p,t) < G(t) then
13          G(t) ← G(p) + dist(p,t)
14          parent(t) ← p
15          if t ∈ open then
16              open.remove(t)
17          end if
18          open.insert(t, G(t), H(t))
19      end if
20  end if
```

Paths found by Theta* are shorter than the obtained by A* or A*PS, but is not guaranteed to find true shortest paths. The main problem in Theta* is that the collision test is frequently performed, which degrades significantly its performance.

3 Heading Changes Measurement and Formulation

In this section we define the computation for the amplitude of heading changes. First we explain the one that could be employed during the search process (for both, heuristic and cost functions) and next, the one to compute the total amount of turns for a path.

To measure the amplitude of a heading change during the search, we define Alpha(p, t, g) (or simply α) as the deviation required to reach a node t starting from the node p and pointing to the node g. Figure 2 shows its graphical representation and Eq. 4 gives the way to compute the value of $\alpha(p, t, g)$ in the range [0°, 180°], being 0° when three nodes are in the same line and ordered $p \rightarrow t \rightarrow g$ or $p \rightarrow g \rightarrow t$; and 180° when they are in the same line and the p node is in the middle, that is, $t \rightarrow p \rightarrow g$ or vice versa.

$$\text{Alpha}(p, t, g) = \arccos \frac{\text{dist}(p, t)^2 + \text{dist}(p, g)^2 - \text{dist}(t, g)^2}{2 \cdot \text{dist}(p, t) \cdot \text{dist}(p, g)} \tag{4}$$

The length of the resulting path is usually employed as a the main measure for the solution optimality when comparing path-planning algorithms. Besides, there are other parameters such as the expanded nodes or the execution time. However, in the literature we cannot find the total amount of heading changes, in terms of total degrees turned. In the case of a mobile robot, for example, the cost of making a turn can be higher than moving forward. The number of heading changes cannot give us enough information about how "smooth" is the path: there could be low number of heading changes, but big in amplitude. In order to select a path-planning algorithm for that case we can take into consideration how this parameter affects the quality of the path. To do that, we consider also a fifth parameter relevant to the study of path-planning algorithms: the *total turn* parameter.

We define the total turn parameter as the sum value of all heading changes (considering that the mobile element is oriented towards the first node of the resultant path) between the start and goal nodes. This is formally expressed in Eq. 5.

$$\text{total turn} = \sum_{i=1}^{n-2} \beta_i \tag{5}$$

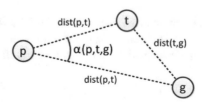

Fig. 2 Graphical representation of $\alpha(p, t, g)$

Each heading change, β_i, is the angle variation produced when we go from the node p_i to the node p_{i+2} through node p_{i+1}. In other words, β_i is the resultant angle of the intersection of a line that crosses the nodes p_i and p_{i+1}, and the line that crosses the nodes p_{i+1} and p_{i+2}. Also, the involved nodes must have the parent relationship: $p_i = \text{parent}(p_{i+1})$ and $p_{i+1} = \text{parent}(p_{i+2})$. We assume that the mobile element can rotate both to the left and to the right, so, in case of the resultant angle β_i is greater than 180°, it must be reduced to obtain an angle in the interval [0°, 180°]. To compute the β_i value we employ the Eq. 6 and for the angle calculation the one shown in Eq. 7, which represents the angle formed by the abscissa axis and the point (x_p, y_p). A visual example of how to compute β_i and the total turn value is shown in Fig. 3.

$$\begin{aligned} \beta_i &= |\text{angle}\,(p_{i+2}, p_{i+1}) - \text{angle}(p_{i+1}, p_i)| \\ \beta_i &- 360 - \beta_i \text{ when } \beta_i > 180 \end{aligned} \tag{6}$$

$$\text{angle}(t, p) = \arccos \frac{(x_t - x_p)^2 + \text{dist}(p, t)^2 - (y_t - y_p)^2}{2 \cdot (x_t - x_p) \cdot \text{dist}(p, t)} \tag{7}$$

This parameter measures the total turn required to reach a path, so we do not get a measure of the amplitude of each turn. We can obtain the mean value for the turn amplitude simply dividing the total turns by the number of heading changes. But it is not really necessary due to the implicit proportionally relationship between these two values: less number of heading changes with higher values of total turns implies high amplitude turns.

Fig. 3 Example of β_i calculation for a small path. β_i expresses the heading between three nodes in the current path. The total turn value is $\beta_1 + \beta_2$, considering that the mobile is pointing to p_2 at the beginning

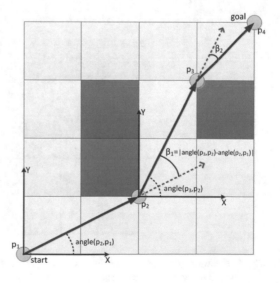

4 Heading Changes as a Heuristic: Efficiency Improvement

Considering the definition of α in Eq. 4, we redefine the heuristic function of a node as the original heuristic employed by the path-planning algorithm plus the value of $\alpha(s, p, g)$, obtaining a new evaluation function, $F(p)$ expressed in Eq. 8.

$$F(p) = G(p) + [H(p) + \alpha(s, p, g)] \tag{8}$$

To compute the α value we require three nodes. One is the current node, p, and the other two are the initial and goal nodes, s and g respectively. Calculating α with these three nodes aims to expand only the nodes that are near (or are contained) in the straight line that connects the start and the goal nodes.

This line is the smallest distance between these two nodes if there are no obstacles blocking the path. For this reason, α takes values in the range [0°, 180°], being 0° when the node belongs to the line and the search algorithm goes towards the goal node. It takes the middle value, 90°, when the deviation of the node is perpendicular to the line. And values greater than 90° implies that reaching the successor node increments the distance to the objective, being the maximum value, 180°, when the node is in opposite direction to the goal.

Figure 4 shows an example with the relevant data for two nodes. The α values for nodes p and p' are 11.31° and 18.43° respectively. If we suppose that the central cell [corresponds to the square formed by nodes (2, 2), (2, 3), (3, 2) and (3, 3)] is blocked, A* expands first the nodes located at the top left of the map, expanding the node p', before the node p. But we can see that the predicted any-angle path length (dotted line) has higher β value for node p' than for node p. The difference in the

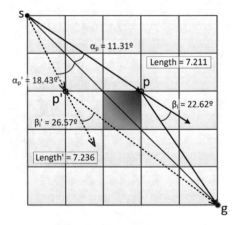

Fig. 4 Example of α values for two nodes. When A* reaches the obstacle in the center, it expands the p' before p. First one is not desirable due to the longer unblocked path length required and the bigger α value. Also, β_i is less for p rather than p'

heading change is 17 % higher for node p' than node p in this example. So expanding the node p' is less likely with our heuristic. Using α with the algorithm forces the search to expand first the node p and relegates nodes far to the optimal unblocked path to the back of the open list.

We can deduce a quick conclusion about applying $\alpha(s, p, g)$ directly to A*: the search algorithm degenerates into a greedy search algorithm which tries to expand only the nodes that belong to the straight line between the start and goal nodes. In Fig. 5 the two represented routes are obtained using the original A* algorithm (blue line) and with the modified heuristic that includes the α value (red line), starting from the top left node and finishing at the bottom right node. Also, the expanded nodes for each search have been marked. Non modified A* expands 41 nodes (empty and red filled circles) and A* with α in its heuristic function expands 14 nodes (red filled circles only).

That is, A* with the modified heuristic expands near 66 % less nodes than the original one. We can also note the tendency of the algorithm to border the obstacles in order to recover the line with $\alpha(s, p, g) = 0$. Therefore, although is predictable that it reduces the number of expanded vertex (and thus, the runtime), the use of α with A* does not imply benefits in terms of path length, and, if there are some obstacles blocking the line between the start and the goal nodes, the number of heading changes could increase. On the other hand, this pseudo-capability to detect obstacles could be useful for any-angle algorithms, so for these reasons, we will employ this heuristic over those kind of path-planning algorithms.

Finally, we must take into consideration that α takes values in the interval $[0°, 180°]$. Considering a map with 100×100 nodes, the cost to transverse from one corner to its opposite corner is $100\sqrt{2} \approx 141$, and we can consider that α is well sized. However, for smaller or bigger maps this shall not be valid. For example, for 50×50 maps, the relative weight of α is double than for a 100×100 map and, for

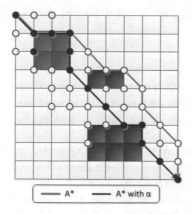

Fig. 5 Path obtained using original A* (*blue*) and A* with the evaluation function expressed in Eq. 8 (*red*). Nodes expanded by A* are represented as a *white circle* whereas the *red filled* are expanded by both. With the modified heuristic, A* expands less nodes and therefore, the runtime is lower

500×500 nodes maps, is the fifth part. In the first case, α has less effect in the search process, so the algorithm tends to behave like the original one, that is, both expand a similar number of nodes; whereas in the second case the heuristic has an excessive cost to expand nodes that are a little bit far from the line between s and g. In order to compensate this fact, we modify the value of α as a function of the map size. This is shown in Eq. 9, taking into consideration a map with $N \times N$ nodes.

$$\alpha(s,p,g) = \alpha(s,p,g) \cdot \frac{N}{100} \qquad (9)$$

But the relative weight of α can modify the behavior of the search algorithm. For this reason, we have considered to multiply its initial value by a c factor, as shown in Eq. 10. As this factor increases, the relative weight of α over the search algorithm grows up, such that the cost of moving away from the line that connects the start and goal nodes is bigger and force the algorithm to expand less nodes. This means that the c factor is inversely proportional to the number of expanded nodes during the search. It must be taken into consideration that, if $c = 0$, the algorithm does not change its behavior (due to the definition of the heuristic as expressed in Eq. 8). So, we can modify the behavior of the search algorithm using c as a parameter. In the experimental section we discuss how the value of α affects the different search algorithms employed, that is, A*PS and Theta*. To summarize, we can say that a high value of α makes the search algorithm greedy, and values near to 0, slightly changes the original behavior. Also, the degradation of path length is directly proportional to the value of the c factor.

$$\alpha(s,p,g) = c \cdot \alpha(s,p,g) \quad \text{with} \quad c \in [0,1] \qquad (10)$$

5   Heading Changes as a Cost Function: S-Theta* Algorithm

We can apply the α value to the cost function of a node, $G(p)$. In the previous section we have employed the nodes p, s (start node) and g (goal node), so the value obtained is fixed for each random node p. If we want to include α as part of the cost function we need to employ other nodes: the heading to reach a node is dependent on the path followed.

Taking into consideration that the cost function is the one that we try to minimize during the search, a good election on the nodes involved in the calculation of α could give us a minimization in the β_i value, and thus, in the total turn of the resultant path. Looking at the formulation of β_i in Eq. 6, we could infer a good relationship between the three nodes: considering the expansion process, we want to reach the goal node g, and, from the current position, be it p, we are trying to expand the node t. The actual direction of the path is the one that follows the line which connects node p with its parent, $q = parent(p)$. So, the deviation to reach the node t from the current position is defined by both, the actual heading and the turn required to go to from p to t having in mind that we want to achieve the goal position.

Finally, we have that the relationship $q = parent(p)$ has important implications: employing the α value as part of the cost function in A* doesn't get any improvement: for A* we have that $q = parent(p) \Rightarrow p \in succ(q)$, so it is possible to locally reduce the zig-zag patterns but with a higher computational cost due to the computation of α. Then, we cannot apply it to A*PS neither.

So, we define $\alpha(q, t, g)$ as the deviation in the trajectory measured from the current position, p, to reach the goal node g through the node t in relation to the current heading defined by the parent of its predecessor $q = parent(p)$ and the node $t \in succ(p)$. $\alpha(q, t, g)$ is represented graphically in Fig. 6.

Taking into consideration the previous issues, we propose the Smooth Theta* (S-Theta*) algorithm that we have developed from Theta*, which aims to reduce the amount of heading changes that the mobile element should perform to reach the goal using the α value as part of the cost function. The evaluation function for the nodes is computed as shown in Eq. 11.

$$F(t) = [G(t) + \alpha(q, t, g)] + H(t) \tag{11}$$

The new term $\alpha(q, t, g)$ (which is added to the accumulated path length) gives us a measure of the deviation from the optimal trajectory to achieve the goal as a function of the direction to follow, conditional to traversing a node t. Considering an environment without obstacles, the optimal path between two points is the straight line. Therefore, applying the triangle inequality, any node that does not belong to that line will involve both, a change in the direction and a longer distance. Therefore, this term causes that nodes far away from that line will not be expanded during the search.

The result is that once the initial direction has changed, the algorithm tries to find the new shortest route between the successor to the current position, t, and the goal node. The shortest route will be, if there are no obstacles, the one with $\alpha(q, t, g) = 0$, i.e., the route in which the successor of the current node belongs to the line

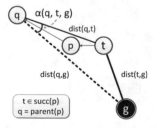

Fig. 6 Graphical representation of $\alpha(q, t, g)$. Actual position is p with $q = parent(p)$. The successor considered is $t \in succ(p)$ and g is the goal node

Fig. 7 Representation of the evolution of $\alpha(t)$. *Arrows* are pointed to the parent of the node after expansion

connecting the parent node of the current position and the goal node. Figure 7 shows how the value of α evolves as the search progresses.

The Algorithm 3 shows the pseudo-code of the *UpdateVertex* function for S-Theta*. α value will be included as a cost in the evaluation function of the nodes, so the algorithm will also discriminate the nodes in the open list depending on the orientation of the search. Thus, a node in the open list may be replaced (which means that its parent will be changed) due to a lower value of α. In contrast, Theta* updates a node depending on the distance to reach it, regardless of its orientation. As a result, the main difference with respect to Theta* is that S-Theta* can produce heading changes at any point, not only at the vertex of the obstacles. This difference can be seen in Fig. 8.

The α value affects the order of the open list, and thus, how the nodes are expanded. To compensate the size of the maps that affects the turn of the mobile element, we have used Eq. 9 explained in the previous section. In the experiments this estimation works fine, so we consider that it is valid. The results show that S-Theta* obtains better values for the total turn parameter than Theta*, with slightly longer path lengths.

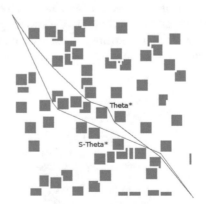

Fig. 8 Resultant paths for Theta* (*red*) and S-Theta* (*blue*) in a random map. Theta* only has heading changes at vertices of blocked cells, while S-Theta* not. Path lengths are 142.28 and 147.82, and total turns 121.54° and 71.56° for Theta* and S-Theta* respectively

---

Algorithm 3 Update vertex function for S-Theta*

```
 1  UpdateVertex(p, t)
 2    α ← α(parent(p),t,g)
 3    if α = 0 or LineOfSight(parent(p),t) then
 4        G_aux ← G(parent(p)) + dist(parent(p),t) + α
 5        if G_aux < G(t) then
 6            G(t) ← G_aux
 7            parent(t) ← parent(p)
 8            if t ∈ open then
 9                open.remove(t)
10            end if
11            open.insert(t,G(t),H(t))
12        end if
13    else
14        G_aux ← G(p) + dist(p,t) + α
15        if G_aux < G(t) then
16            G(t) ← G_aux
17            parent(t) ← p
18            if t ∈ open then
19                open.remove(t)
20            end if
21            open.insert(t,G(t),H(t))
22        end if
23    end if
```

5.1 Implementation Issues

We suggest two procedures to improve the implementation of the S-Theta* algorithm. The α computation degrades the S-Theta* performance respect Theta* because of both, the cost of floating point operations and the increase of the cost of checking the lines of sight (to make less heading changes, the algorithm needs to check the line of sight for bigger map sections). However, the degradation is not very significant in terms of CPU time as we will see in the experimental section. This is thanks to the optimization procedures that will be outlined here and the fact that S-Theta* expands less nodes than the original Theta*.

Procedure 1 The distance between the node t and the goal node ($dist(t, g)$) is static and we can save many operations if once it has been calculated is stored as a property of the node. Before computing the distance between a node and the goal node we will check if this operation has already been performed, so we can save this computation. We only need to initialize this data to a negative value in the instance of the nodes prior to the search.

Procedure 2 If $\alpha(q, t, g) = 0$ means that the node and the predecessor of its parent are in the same line, i.e. the nodes $q = parent(p)$, p and t are in the same line. Therefore, and given that $t \in succ(p)$ and being the set $succ(p)$ the neighbours reachable from p, it follows that there is a line of sight between t and q. This saves the line of sight checking.

6 Experimental Results

In this section we show the results obtained by comparing the A*, A*PS, Theta* and S-Theta* algorithms. We also include A*PS and Theta* using the modified heuristic based on the Alpha computation, with a weight factor (c) of 1. The following subsections show the results obtained by running the algorithms on outdoors (random obstacles) maps and indoor maps with interconnected rooms and corridors. For both cases, we have taken into consideration the average values for the following parameters: (i) the length of the path, (ii) the total number of accumulated degrees by the heading changes (total turn), (iii) the number of expanded nodes during the search, and, (iv) the CPU time or search run-time. We do not present here the number of heading changes, since we think that the value of the total turn provides more useful information.[1]

The design of the test-bench guarantees that all problems have at least one solution, and the different path-planning algorithms use the same method and structures to manage the grid information. The execution is done on a 2.8 GHz Intel Core i5 with 4 GB of RAM under Ubuntu 11.04.

6.1 Outdoor Maps

Figure 9 shows the results obtained for the resolution of 5000 random maps of 700×700 nodes, gradually increasing the percentages of blocked cells to 5, 10, 20, 30 and 40 % (each obstacle group has 1000 maps). The way to generate the maps guarantee that there will be at least one valid path from any starting point to the goal. To do that, each time an obstacle is randomly introduced, we force that around the obstacle there are free cells and these free cells cannot overlap with another obstacle. In all cases the initial position corresponds to the coordinates (0, 0) and the objective is to reach a node in the last column randomly chosen the row from the bottom fifth (699, 560–699).

As we can see, the algorithm that obtains the shorter routes is Theta*, then all algorithms that use Alpha have better path length than without it. A* obtains the worse performance. From the modified algorithms, there is one that obtains closer path lengths to A* when the number of blocked cells increases: S-Theta. For 40 %

---

[1]Datasets and full resolution bar plots could be obtained in: http://ogate.atspace.eu/novelappai/dataset.html.

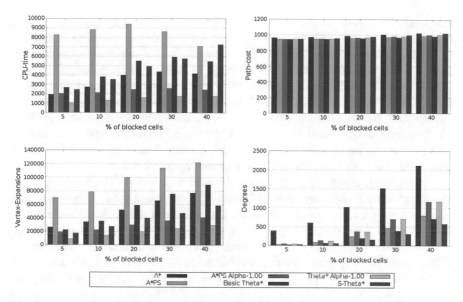

Fig. 9 Results for the execution of different path-planning algorithms over 5000 random generated maps (each obstacle group has 1000 maps). From *top left* to *bottom right* run-time in milliseconds, path length, number of expanded vertex, total turn in degrees

of blocked cells, A* obtains a path length of 1027.80 and S-Theta* gets 1026.84. This implies that S-Theta has an important degradation in the path length respect to its former (Theta*) that grows from 0.57 with 5 % of obstacles to 4.22 with 40 % of blocked cells. Respect to the algorithms that employ Alpha as part of the heuristic function, we obtain less degradation than the one obtained in S-Theta* for both cases, A*PS Alpha and Theta* Alpha. It is visible that the degradation is directly dependent on the number of blocked cells, being more remarkable with higher number of obstacles. Also, for A*PS and Theta* using Alpha as part of the heuristic function, the degradation is also proportional to the value of the c factor: small values such as 0.25 has little effect in the path-planning algorithm, and the greater degradation corresponds to the value presented here, $c = 1.00$. More detailed results of how the parameters evolve as a function of Alpha could be checked in Muñoz and R-Moreno [15].

For the total turn parameter (the total amount of heading changes measured in degrees) we obtain that S-Theta* has the best values (except for the minimum obstacles case) and, obviously, A* is the worst due to the constraint of heading changes to multiples of 45°. Here we can found that the difference in the total turns between Theta* and S-Theta* is bigger with more obstacles. Theta* requires 2.12 % degrees more than S-Theta* for maps with 10 % of obstacles and 22.73 % degrees more when the blocks cells grow up to 40 %. This difference shows that S-Theta* tends to maintain smoothed paths at the expense of the path length. For A*PS and Theta* with Alpha as part of the heuristic function we obtain the same behavior of the path length: turn performance, as measured by heading change, degrades as a function of both the percentage of obstacles and the value of c factor.

To get a more accurate analysis of the optimality of the solution for the different path-planning algorithms, in Fig. 10 we graph the sum of the path length and total turn against the percentage of blocked cells. We can observe that the worst result is for A*, followed by A*PS Alpha and Theta* Alpha (whose results are practically superposed). In the fourth position there is A*PS. Then, for 5 % and 10 % of blocked cells the last two algorithms, Theta* and S-Theta*, have very similar results, but for more obstacles, S-Theta* obtains better values, being more remarkable when the number of obstacles grow up to 40 %.

In the case of the run-time, the best values are achieved by Theta* using Alpha as part of the heuristic function. A*PS and Theta* using Alpha improves their performance in a factor close to 3 for the first algorithm and 2 for the second one. The worst run-time corresponds to the original version of A*PS, it requires to run A* and then post process it. However, both A*PS Alpha-1.00 and Theta* Alpha-1.00, have better run-time than A* in all cases (with the exception of A*PS Alpha-1.00 and 5 % of blocked cells). Also we can observe that S-Theta* requires less time than Theta*, except for the higher number of obstacles, when it requires significantly more time (more than 1 s on average).

Finally, and directly proportional to the run-time, there is the number of expanded vertex. We only want to remark the difference between the expanded vertex by A* and A*PS. This is because A*PS uses the euclidean distance instead of the octile one as A* does. Also, we can see that S-Theta* expands significantly less vertex than the former one, Theta*, but, the run-time for the two algorithms are similar: the reason for this is the computational effort required to perform the Alpha computation and, as consequence of the minimization of the heading changes, S-Theta* requires to check the line of sight for longer segments, degrading more its performance as consequence of the line of sight check.

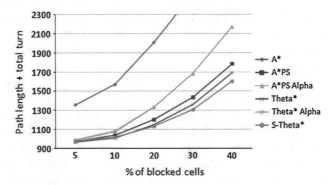

Fig. 10 Evolution of path length + total turn respect to the percentage of blocked cells for 5000 random generated maps

6.2 Indoor Maps

For indoor maps, we have run the algorithms over 900 maps with different sizes: 300 × 300, 450 × 450 and 600 × 600 nodes (300 maps per size), always starting from the upper left corner (0, 0) and reaching the target in the opposite corner. The indoor maps are generated from the random combination of square patterns that represent different configurations of corridors and rooms. These patterns are designed in a way that we can access to the next pattern through doors on each side, symmetrically placed. Figure 11 shows the results for the four comparison criteria.

The data obtained for indoor maps are similar, in general terms, to those obtained in random maps. In all cases the path length is shorter in Theta* than the rest of algorithms, although in this case A* obtains better results than S-Theta*. Furthermore, the path degradation in S-Theta* respect Theta* is 6.78 % higher for the smallest size maps, rising up to 8.73 % in the 600 × 600 nodes maps. For A*PS Alpha and Theta* Alpha the path length is slightly longer than the former ones, but always less than the obtained with A* and S-Theta*.

The results for the total turns show that, again, S-Theta* has the best values, being A* the worst. Also, we can appreciate that A*PS obtains better values than Theta*. Comparing this one with S-Theta*, we obtain that Theta* requires 19 % degrees more than S-Theta* for small maps and 25.44 % degrees more for the bigger ones. In the case of the algorithms with the Alpha value in its heuristic, the

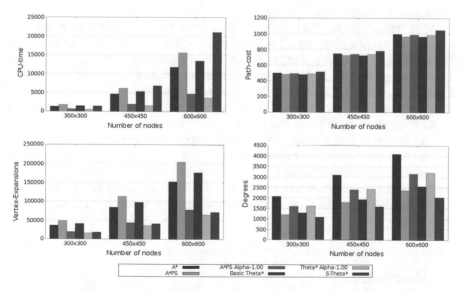

Fig. 11 Results for the execution of different path-planning algorithms over 900 indoor generated maps. Each group (300 × 300, 450 × 450 and 600 × 600 nodes) has 300 maps. From *top left* to *bottom right* run-time in milliseconds, path length, number of expanded vertex, total turn in degrees

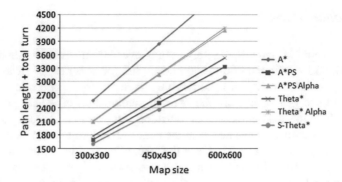

Fig. 12 Evolution of path length + total turn respect to the map size for 900 indoor generated maps

degradation of the total turn is, once again, proportional to the c factor, and in all cases, the results obtained are better than the achieved with A*.

In Fig. 12 we present the value for the couple of path length and total turn for every algorithm and map size. We obtain the following order in performance, from worse to better: A*, A*PS Alpha and Theta* Alpha (again superimposed), Theta*, A*PS and S-Theta*. We can see that Theta* obtains worse results than A*PS, and, the difference between Theta* and S-Theta* is greater than the one in the random maps. This is due to the huge difference in total turn and the optimization of that parameter performed by S-Theta*.

With respect to the run-time we observe that there is a big difference as a function of the size of the map. For small maps A*PS is the one that requires more time to achieve a solution. But for medium and bigger maps, S-Theta* obtains the longer run-times. Also, in these maps, A* gets better values than A*PS and Theta*. Finally, in all maps, the best run-time corresponds to Theta* Alpha, followed by A*PS Alpha. Using the modified heuristic improves the performance of the two former algorithms by a factor close to 3.

The reason to this speedup resides in the number of expanded vertex, A*PS Alpha and Theta* Alpha expand near one half of the number of expanded vertex of the original algorithms. We see that S-Theta* expands less vertex than Theta* but obtains worse run-times. Again, this is due to the computational cost of the Alpha calculation and the necessity of performing the line of sight for bigger areas than the former one.

7   Conclusions

In this chapter, we have presented a new way to improve any-angle path-planning algorithms. To do this, we have introduced the angle deviation between three nodes. Applying this parameter in the evaluation function of the search process we can modify the algorithms in two ways: (i) improving its efficiency when we use it as a dependent domain heuristic with a slight degradation of path length and total turn of the generated paths, and, (ii) obtaining a new algorithm, which we have called S-Theta*, that minimizes the heading changes during the search process, not only the path length as the former one do.

The angle deviation parameter, that we have called it Alpha, has three advantages: first, it is easy to implement, and is valid for any path-planning algorithm based on A*. Second, we can modify the behavior of Alpha using a factor (known as c factor in the paper) to deal between the run-time and the degradation of the other parameters. Finally, Alpha affects to the required memory: using this value makes that the search algorithm expands less nodes, and thus, less memory usage.

In this way, to test the advantages of the proposed innovation, we have performed an experimental setting to test A*, A*PS, Theta* and the same algorithms including Alpha in their evaluation function: A*PS Alpha and Theta* Alpha, which use the term as part of the heuristic function; and S-Theta* that considers the heading change as part of the cost function.

As a conclusion, we can say that using Alpha in A*PS achieves better results in terms of number of expanded nodes (and thus, run-time) but degrades the path length proportionally to both the c factor and the number of blocked cells. However, for small c values the degradation is not very high and, using $c = 0.50$ the results show that A*PS gets better results in all parameters than A* with less run-time. The same can be applied to Theta*, taking into consideration that the degradation is more remarkable.

Also, we have presented S-Theta*, an algorithm that is based on Theta* and that employs the Alpha value as part of the cost function. The new algorithm effectively reduces the number of heading changes required by the search algorithm to achieve the objective, as well as the total cost associated with these turns. As the experimental results show, the S-Theta* algorithm improves the original algorithm Theta* on the total turn, in exchange of a slight degradation on the length of the path. But, taking into consideration that an optimal solution is the one that includes both the length of the path and the total turn, S-Theta* gets better results than Theta*.

Finally, with the result obtained for the couple path length and total turn parameters, we can observe that these values are dependent of both, the size of the map and the number of obstacles, but in different manners: the dependence is lineal as a function of the size of the map, and the grow of these parameters is exponential as a function of the number of blocked cells.

Acknowledgements Pablo Muñoz is supported by the European Space Agency (ESA) under the Networking and Partnering Initiative (NPI) *Cooperative systems for autonomous exploration missions*. This work was supported by the Spanish Ministry of Economy and Competitiveness under the project TIN2014-56494-C4-4-P and the Junta de Comunidades de Castilla-La Mancha project PEII-2014-015-A.

References

1. S. Russell, P. Norvig, *Artificial Intelligence: a Modern Approach* (Morgan Kaufmann Publishers, 3 edn., 2009)
2. D. Higgins, *AI Game Programming Wisdom*, chapter 3.3 (Charles River Media, 2002)
3. E. Dijkstra, A note on two problems in connexion with graphs. Numer. Math. **1**, 269–271 (1959)
4. P. Hart, N. Nilsson, B. Raphael, A formal basis for the heuristic determination of minimum cost paths. IEEE Transac. Syst. Sci. Cybern. **4**, 100–107 (1968)
5. I. Millington, J. Funge, *Artificial Intelligence for Games* (Morgan Kaufmann Publishers, 2 edn., 2009)
6. A. Botea, M. Muller, J. Schaeffer, Near optimal hierarchical path-finding. J. Game Dev. **1**, 1–22 (2004)
7. D. Ferguson, A. Stentz. Field D*: an interpolation-based path planner and replanner, in *Proceedings of the International Symposium on Robotics Research (ISRR)* (October 2005)
8. G. Ayorkor, A. Stentz, M. B. Dias, Continuous-field path planning with constrained path-dependent state variables, in *ICRA 2008 Workshop on Path Planning on Costmaps* (May 2008)
9. A. Nash, K. Daniel, S. Koenig, A. Felner. Theta*: any-angle path planning on grids, in *Proceedings of the AAAI Conference on Artificial Intelligence (AAAI)*, pp. 1177–1183 (2007)
10. K. Daniel, A. Nash, S. Koenig, A. Felner, Theta*: any-angle path planning on grids. J. Artif. Intell. Res. **39**, 533–579 (2010)
11. S. Choi, J.Y. Lee, and W. Yu. Fast any-angle path planning on grid maps with non-collision pruning. In *IEEE International Conference on Robotics and Biomimetics*, pages 1051–1056, Tianjin, China, December 2010
12. P. Yap, Grid-based path-finding, in *Advances in Artificial Intelligence*, vol. 2338 of *Lecture Notes in Computer Science*, pp. 44–55 (Springer, Berlin, 2002)
13. M. Kanehara, S. Kagami, J. Kuffner, S. Thompson, H. Mizoguhi. Path shortening and smoothing of grid-based path planning with consideration of obstacles, in *IEEE International Conference on Systems, Man and Cybernetics, ISIC,* pp. 991–996 (October 2007)
14. C. E. Thorpe, L.H. Matthies. Path relaxation: path planning for a mobile robot, in *OCEANS Conference*, pp. 576–581 (September 1984)
15. P. Muñoz, M. D. R-Moreno, Improving efficiency in any-angle path-planning algorithms, in *6th IEEE International Conference on Intelligent Systems (IEEE-IS)*, pp. 213–218 (Sofia, Bulgaria, September 2012)

C × K-Nearest Neighbor Classification with Ordered Weighted Averaging Distance

Gozde Ulutagay and Efendi Nasibov

Abstract In this study, OWA (Ordered Weighted Averaging) distance based $C \times K$-nearest neighbor algorithm ($C \times K$-NN) is considered. In this approach, from each class, where the number of classes is C, K-nearest neighbors are taken. The distance between the new sample and its K-nearest set is determined based on the OWA operator. It is shown that by adjusting the weights of the OWA operator, it is possible to obtain the results of various clustering strategies like single-linkage, complete-linkage, average-linkage, etc.

1 Introduction

Classification of objects is an important field of research and it has an increasing popularity in many application areas including artificial intelligence, statistics, pattern recognition, and medicine [1, 17]. Obviously, as the a priori knowledge about the problem domain increases, the power of the classification algorithm to reflect the actual situation increases. In many pattern recognition problems, the classification of an input pattern is based on data where the respective sample sizes

G. Ulutagay (✉)
Department of Industrial Engineering, Izmir University, Gursel Aksel Blv 14, 35350 Izmir, Turkey
e-mail: gozde.ulutagay@izmir.edu.tr; gozde.ulutagay@gmail.com

E. Nasibov
Department of Computer Science, Dokuz Eylul University, Tinaztepe Campus, 35160 Izmir, Turkey
e-mail: efendi.nasibov@deu.edu.tr

E. Nasibov
Institute of Cybernetics Azerbaijan National Academy of Sciences, Baku, Azerbaijan

© Springer International Publishing Switzerland 2016
M. Hadjiski et al. (eds.), *Novel Applications of Intelligent Systems*,
Studies in Computational Intelligence 586, DOI 10.1007/978-3-319-14194-7_6

of each class are small and possibly not representative of the actual probability distributions, even if they are known [4, 17]. In such cases where there is a lack of such information, many algorithms make use of distance or similarity among samples as a means of classification.

Nearest neighbor is one of the most important algorithms of data mining technology. Since its proposition, K-nearest neighbor has been investigated and improved by many researchers [2, 3, 5, 8]. The main idea of these algorithms except study [15] is to evaluate all the nearest points from different classes within a certain neighborhood of an unclassified point together. The unclassified point is assigned to the dominant class. However, in $C \times K$-NN studies, elements from all of the classes are evaluated and the point is assigned to a class according to the smallest aggregated distance. In the evaluation process, min and max strategies are handled.

The main idea of this study is to evaluate the proximity of neighbors by using a more general aggregation method OWA. For the K-nearest neighbors of each class, by changing the weighting scheme different classification results, which coincide with the well-known methods such as single-linkage, complete-linkage, average-distance, can be obtained.

The rest of the paper is organized as follows: In the next section, the well-known K-NN algorithm is reviewed. In Sects. 3, 4 and 5, Fuzzy K-NN algorithm, Weighted K-NN algorithm and Optimally Weighted Fuzzy K-NN algorithm are described. In Sect. 6, single, complete, and average linkage distances are explained. $C \times K$-NN algorithm is handled in Sect. 7. In the next section Ordered Weighted Averaging Distance is described. OWA-Distance Based $C \times K$-NN algorithm, which is the main subject of this study, is described in detail in Sect. 9 and a numerical example for this algorithm is given in Sect. 10. The study is concluded by the last section.

2 K-Nearest Neighbor Algorithm

Cover and Hart [1] proposed K-nearest neighbor algorithm which is a non-parametric approach that represents one of the simplest and most intuitive methods in the field of statistical discrimination. In this well-known algorithm, a new data is put into the closest class in the learning set, with respect to the covariates used. The similarity is determined by using distance measures.

Let $L = \{(y_i, x_i), i = 1,\ldots, n\}$ be a learning set of labeled samples, where $y_i \in \{1,\ldots, c\}$ denote class labels and the vector $x_i' = (x_{i1},\ldots, x_{ip})$ represents the predictor values. The determination of nearest neighbors is based on an arbitrary distance function. Then for a new sample (y, x), the nearest K-neighbor set is determined and the majority class of the nearest neighbor is selected as prediction for y. The pseudo-code of the K-Nearest Neighbors algorithm is given below:

```
BEGIN
   Input  x of unclassified data.
      Set  K, 1≤ K ≤ n  and   i =1 .
   DO UNTIL ( K -nearest neighbor found)
      Calculate the distance between  x and  x_i
      IF ( i ≤ K ) THEN Assign  x_i  to the set of  K -nearest neighbors
      ELSE IF ( x_i  is closer to  y  than any other previous neighbor)
THEN
            Delete the farthest sample in the set of K -nearest neighbors.
            Assign  x_i  to the set of  K -nearest neighbors.
      END IF
      i = i +1
      END DO UNTIL
      Mark the majority class represented in the set of  K -nearest
      neighbors.
      IF (a tie exists) THEN
            Calculate the sum of distances of neighbors in each class tied.
            IF (no tie occurs) THEN
                Classify  x  in the class of minimum sum.
            ELSE
                Classify  x  in the class of last minimum found.
            END IF
      ELSE
            Classify  x  in the majority class.
      END IF
   END
```

3   Fuzzy *K*-Nearest Neighbor Algorithm

Keller et al. [9] proposed the fuzzy version of the *K*-nearest neighbor algorithm which assigns class membership to a sample vector rather than assigning the vector to a certain class. In Fuzzy *K*-nearest neighbor algorithm, arbitrary assignments are not made as well as the vector's membership values should provide a level of assurance to accompany the resultant classification.

Let $X = \{x_1, x_2, ..., x_n\}$ be a set of n labeled samples, $\mu_i(x)$ be the assigned membership of the vector x, and $\mu_{ij}(x)$ be the membership degree of the jth vector of the labeled sample set to the ith class given as follows:

$$\mu_i(x) = \frac{\sum_{j=1}^{K} \mu_{ij}\left(1/\|x - x_j\|^{2/(m-1)}\right)}{\sum_{j=1}^{K}\left(1/\|x - x_j\|^{2/(m-1)}\right)} \tag{1}$$

where K is the number of the nearest neighbors. The pseudo-code of the Fuzzy K-Nearest Neighbors Algorithm is given below:

```
BEGIN
Input x of unclassified data.
    Set K, 1 ≤ K ≤ n
    Set i = 1 .
    DO UNTIL ( K -nearest neighbor found)
        Calculate the distance between x and x_i
        IF ( i ≤ K ) THEN
            Assign x_i to the set of K -nearest neighbors
          ELSE IF ( x_i is closer to x than any other previous neighbor
THEN
            Delete the furthest sample in the set of K -nearest neighbors.
            Assign x_i to the set of K -nearest neighbors.
        END IF
        END DO UNTIL
        FOR EACH CLASS i
            Calculate the membership degree μ_i(x) according to the formula
            (1).
        END FOR
    END
```

4   Weighted K-Nearest Neighbor Algorithm

Paik and Yang [12] proposed to use combinations of various K-nearest neighbor classifiers by using different value of K and different subsets of covariates for the sake of improving the results of one single K-nearest neighbor prediction. Since the method known as adaptive classification by mixing (ACM), it is also suitable for working with weights. The difference is that rather than giving weights to the samples, a weighting scheme for all of the classifiers is calculated according to their classification probabilities [14].

Let $X_1, X_2,\ldots, X_n$ denote a sample size of n from a random variable with density f. The kernel estimate of f at the point x is as follows:

$$\hat{f}_h(x) = \frac{1}{nh} \sum_{i=1}^{n} K\left(\frac{x - X_i}{h}\right)$$

(2)

which is also known as Parzen window. The transformation from distance to weights is performed by using a kernel function $K(\cdot)$ which has the following properties:

(i) $K(d) \geq 0$ for all $d \in R$;
(ii) $K(d)$ reaches its maximum for $d = 0$;
(iii) $K(d)$ is monotonic descending for $d \rightarrow \pm\infty$.

Some of the widely known kernel functions are given in Figs. 1 and 2 and Table 1.

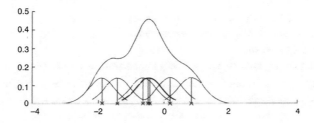

Fig. 1 Kernel estimate that shows individual kernels [14]

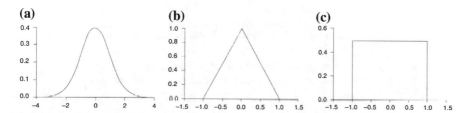

Fig. 2 **a** Gaussian, **b** triangular, and **c** rectangular kernel functions

Table 1 Some of the widely used kernel functions	Kernel type	Formula
	Gaussian	$K(d) = \frac{1}{\sqrt{2\pi}} e^{-\frac{d^2}{2}}$
	Triangular	$K(d) = \begin{cases} 1 - |d|, & \text{if } |d| \leq 1 \\ 0, & \text{elsewhere} \end{cases}$
	Rectangular	$K(d) = \begin{cases} \frac{1}{2}, & \text{if } |d| \leq 1 \\ 0, & \text{elsewhere} \end{cases}$

The pseudo-code of the Weighted *K*-Nearest Neighbors Algorithm is given below:

```
BEGIN
   Input  x of unclassified data.
   Set  K, 1 ≤ K ≤ n -1
   Fix a kernel function K(.);
   Set i = 1.
   DO UNTIL ( K +1-nearest neighbor found)
     Calculate the distance between x and x_i
     IF ( i ≤ K +1) THEN
          Assign  x_i  to the set of  K +1-nearest neighbors
     ELSE IF ( x_i  is closer to  x  than any other previous neighbor) THEN
          Delete the furthest sample in the set of K -nearest neighbors.
          Assign  x_i  to the set of  K -nearest neighbors.
     END IF
     Calculate the normalized distance D_i=D(x,x_i) according to the
     formula given below:
```

$$D_{(i)} = D(x, x_{(i)}) = \frac{d(x, x_{(i)})}{d(x, x_{(k+1)})}$$

```
     Transform the normalized distance D(x,x_i) into weights w_i=K(D_i)
     with the any kernel function K(.) according to the formula given
     in Table 1.
   END DO UNTIL
   Mark the class with majority sum of weights according to the
   formula given below represented in the set of  K -nearest
   neigbors:
```

$$\hat{y} = \max_r \left(\sum_{i=1}^{k} w_{(i)} I(y_{(i)} = r) \right)$$

```
   IF (a tie exists) THEN
      Classify  x  in the last found class with majority sum of
      weights.
   ELSE
      Classify  x  in the majority class.
   END IF
END
```

5 Optimally Weighted *K*-Nearest Neighbor Algorithm

Among the simplest and the most intuitively appealing classes of non-probabilistic classification procedures are those that weight the evidence of nearby sample observations most heavily. More specifically, one might wish to weigh the evidence of a neighbor close to an unclassified observation more heavily than the evidence of another neighbor which is at a greater distance from the unclassified observation.

In order to overcome the distance choice dependency problem of fuzzy *K*-NN algorithm, Pham [13] proposed a computational scheme to obtain the optimal weighting coefficients in terms of statistical measure and combine these weights with various degree memberships for classification purpose by the fuzzy K-NN algorithm which is called optimally weighted fuzzy K-nearest neighbor algorithm (OWFKNN).

For an unknown data x_u, OWFKNN algorithm designates a membership degree to a class label y as an optimally weighted linear combination of the membership degrees of k-nearest samples:

$$\mu_{yu} = \sum_{i=1}^{k} w_i \mu_{yi} \tag{3}$$

Note that in the expression given above, μ_{yu} and μ_{yi} are defined according to the Eq. (1) similar to μ_{ij} and $\mu_i(x)$, respectively, and w_i, $i = 1,\ldots, k$ are the set optimal weights that could be derived from the estimated value $\hat{x}_u$ of the unknown sample x_u and show the relationship between x_i and x_u:

$$\hat{x}_u = \sum_{i=1}^{k} w_i x_i \tag{4}$$

where $x_i,\ldots, x_k$ are the current data. Equation (1) can be simplified as follows:

$$\mu_{yu} = \frac{\sum_{i=1}^{k} c_i \mu_{yi}}{\sum_{i=1}^{k} c_i} \tag{5}$$

where

$$c_i = 1/\|x - x_j\|^{2/(m-1)} \tag{6}$$

Although there are various approaches to obtain the weights of neighbor data, one of the well-known method is to minimize the average error of estimation. If r_j is the error between the estimated and the actual value of x_j, then the average error of k estimates can be defined:

$$r_{average} = \frac{\sum_{j=1}^{k} r_j}{k} \tag{7}$$

where

$$r_j = \hat{x}_j - x_j. \tag{8}$$

Since this minimization procedure is not realistic due to unknown actual values of x_i, the best linear unbiased estimator is kriging which handles the unknown values as the outcome of a random process and treats the problem by statistical techniques. Alias, the difference of the random variables modeling the estimate and the actual value, i.e. the variance of the modeled error, could be minimized whereas the minimization of the variance of actual errors is not realistic. The set of optimal weights via kriging is calculated by solving the system of equations given below:

$$Cw = D \tag{9}$$

where

$$C = \begin{bmatrix} C_{11} & \cdots & C_{1k} & 1 \\ \cdot & \cdots & \cdot & \cdot \\ \cdot & \cdots & \cdot & \cdot \\ \cdot & \cdots & \cdot & \cdot \\ C_{k1} & \cdots & C_{kk} & 1 \\ 1 & \cdots & 1 & 0 \end{bmatrix} \tag{10}$$

$$w = \begin{bmatrix} w_1 & \cdots & w_k & \beta \end{bmatrix}^T \tag{11}$$

$$D = \begin{bmatrix} C_{1u} \ldots C_{ku} 1 \end{bmatrix}^T \tag{12}$$

where C_{ij} is the covariance between x_i and x_j, w_i, $i = 1,\ldots, k$ are the optimal, i.e. kriging weights, and β is the Lagrange multiplier. Then the optimal weights are acquired by solving the following system

$$w = C^{-1}D \tag{13}$$

where C^{-1} is the inverse of the covariance matrix. Then, following equation system is achieved by fitting the variables in kriging system:

$$C_{iu} = \sum_{j=1}^{k} w_j C_{ij} + \beta \mathrm{I}, \quad \text{for } \forall i = 1, \ldots, k. \tag{14}$$

Note that in order to avoid negative weights of the kriging system and make the estimation robust, the following normalization can be used where w_i^* is the corrected weight and $\alpha = - \min_i w_i$:

$$w_i^* = \frac{w_i + \alpha}{\sum_{i=}^{k} (w_i + \alpha)}, \quad \text{for } \forall i. \tag{15}$$

Finally, μ_{yu}, i.e. the fuzzy membership degree of the unknown data sample x_u to class label y, is computed by using Eq. (3).

The above-mentioned methods in Sects. 2, 3, 4 and 5 are based on K-nearest neighbor algorithm. This study has a tendency to perform classification similar to the viewpoint of study, Ulutagay and Nasibov [15]. Therefore, it will be suitable to mention some linkage concepts in the following section.

6 Linkage Distances

In hierarchical clustering algorithms, many inter-cluster distance approaches are of concern. They function by either in a process of successive merges or in a process of successive divisions. Agglomerative hierarchical methods start proceeding with the individual objects. At each step, the most similar objects are put into the same cluster, and these initial groups are merged according to their similarities. On the other hand, divisive hierarchical methods work conversely. At each step group of objects is divided into two subgroups such that the objects in the one subgroup are far from the objects in the other. These subgroups are then further divided into similar subgroups, and this process continues until there are as many subgroups as objects, i.e. until each object forms a group [20]. In study [15], inter-cluster linkage approaches used in hierarchical clustering to measure the distance between the classified point and its nearest neighbor points' class was of interest.

6.1 Single-Linkage Distance

Distances or similarities between pairs of objects are the inputs for the single linkage algorithm. By merging nearest neighbors, groups are composed from the individual objects [20]. Let $Dist(A, B)$ be the distance between clusters A and B, and y_i and z_j be the elements of clusters A and B, respectively. Then the single-linkage method defines the inter-cluster distance as the distance between the elements in each of the two clusters that are nearest:

$$Dist(A, B) = \min_{y_i \in A, \, z_j \in B} d(y_i, z_j). \tag{16}$$

The clusters formed by the single-linkage method will be unchanged by any assignment of distance or similarity that gives the same relative orderings as the initial distances.

6.2 Complete-Linkage Distance

Complete-linkage approach is analogous to single-linkage. However, the distance or similarity between clusters is measured by the distance between the two elements from different clusters that are most distant. In complete-linkage method, the inter-cluster distance is defined as the distance between the elements in each of the two clusters that are most distant:

$$Dist(A, B) = \max_{y_i \in A, \, z_j \in B} d(y_i, z_j). \tag{17}$$

Fig. 3 Illustration of the
proceeding principle of
a single, **b** complete, and
c average linkage methods

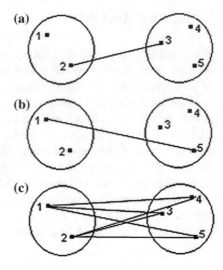

Hence, complete-linkage guarantees that all the objects in the same cluster are within some maximum distance or minimum similarity of each other.

6.3   Average-Linkage Distance

Average-linkage treats the distance between two clusters as the average distance between all pairs of items where one member of a pair belongs to each cluster:

$$Dist(A, B) = \frac{1}{n_a} \frac{1}{n_b} \sum_{y_i \in A} \sum_{z_j \in B} d(y_i, z_j). \tag{18}$$

For average-linkage clustering, changes in the assignment of distances or similarities can affect the arrangement of the final configuration of clusters, even though the changes preserve relative orderings.

An illustration of these three linkage approaches is given in Fig. 3.

In this study, OWA-linkage distance which can be considered as a generalization of the above-mentioned linkage distances. The detailed information will be given in the proceeded sections.

7 $C \times K$-Nearest Neighbor Algorithm

Ulutagay and Nasibov [15] proposed $C \times K$-nearest neighbor algorithm which considers the effect of all the nearest neighbors from each class. The difference of $C \times K$-nearest neighbor algorithm from the classical K-nearest neighbor algorithm is in the total number of nearest neighbors considered [15]. In the classical approach, the distance between each sample and the new data is calculated, and the smallest K of all these distances are taken into account. Among these distances, the most dominating class, i.e. either the class which contains maximum number of nearest neighbors or the closest sample to the new data if the number of nearest neighbor is equal in each class, is determined and the new sample is assigned to the dominating class. However, in $C \times K$-nearest neighbor algorithm, totally $C \times K$ samples are considered, where C is the number of classes, and K is the number of nearest samples selected from each class. By taking account K neighbors, the distance between the new sample and the class is calculated. Finally, the new sample is assigned to the nearest class.

Let $X = \{x_1, x_2, \ldots, x_n\}$ be a set of n labeled samples, $C_j = \{x_1^j, x_2^j, \ldots, x_{n_j}^j\}$ $j = 1$, $2, \ldots C$, be the a priori known classes, n_j be the number of elements in class C_j where $n_1 + n_2 + \cdots + n_c = n$. The pseudo-code of $C \times K$-Nearest Neighbors Algorithm is given below:

```
BEGIN
    Input  x of unclassified data.
    Set  K, 1≤ K ≤ n
  FOR EACH CLASS  C_j
  DO UNTIL ( K -nearest neighbor found in class C_j )
    Set i = 1 .
    Calculate the distance between  x and  x_i^C
    IF ( i≤ K ) THEN
        Assign  x_i^{C_j} to the set of  K -nearest neighbors in C_j
      ELSE IF ( x_i^{C_j} is closer to  y than any other previous neighbor) THEN
        Delete the farthest sample in the set of K -nearest neighbors.
        Assign  x_i^{C_j} to the set of  K -nearest neighbors in C_j.
      END IF
      i = i + 1
      END DO UNTIL
      Calculate the average distance d_j of x from  K -nearest neighbors
      of class  C_j
  END FOR
  Mark the class with minimum distance d_r=min_j d_j;
  Classify  x  in the class r of the last minimum found.
  END
```

8   Ordered Weighted Averaging OWA Distance

Yager [19] proposed Ordered Weighted Averaging (OWA) method as a way for providing aggregations which lie between *max* and *min* operators and also which is one the best formulas in order to determine the average representative of a finite number set. In essence, OWA is the weighted aggregation of the set of elements. Therewithal, the weights are assigned to the ranking positions rather than to elements themselves. Such a situation enables to perform the aggregation strategy according to the pessimism or optimism degree of the decision-maker. Since its proposal, OWA aggregation operator has been studied enthusiastically and applied to numerous fields. Let the set of real numbers $A = \{a_1, a_2,..., a_n\} \equiv \{a_i\}_{i=1}^{n}$ be determined.

Definition 1

An OWA operator of n dimension is a mapping $f{:}R^n \rightarrow R$ with weighting vector $w = \{w_1, w_2,..., w_n\}$ and is determined as follows:

$$OWA_w(a_1, a_2, ..., a_n) = \sum_{i=1}^{n} w_i a_{(i)} \qquad (19)$$

where $a_{(i)}$ is the ith largest value among the elements $a_1, a_2,..., a_n$, and $w = (w_1, w_2,..., w_n)$ is the weighting vector satisfying the following conditions:

(i) $w_i \in [0, 1], \quad i = 1,...,n$
(ii) $\sum_{i=1}^{n} w_i = 1$

Note that the weighting vector stands for the averaging strategy independent of the particular values of the averaged elements. Different aggregation strategies depend on the choice of the weighting vector. OWA operator transforms into widely used operators simply by using various weighting vectors [11, 19].

There are many approaches to determine the weights and their characteristics. For example, in study [18], normal probability density function is used to generate positional weights. Mainly two measures characterizing the weighting vector are used. The first one is the "orness", which measures the degree of agreement with the logical "or" and the second measure is the measure of dispersion of the aggregation or entropy [20]. Sometimes it is of interest to solve the inverse problem. Hence, one can be interested in calculating the weights when either orness or entropy values are given. For instance, Fuller and Majlender [7] interested in determining the associated weights which provide maximum entropy with respect to the given orness value by transforming the Yager's OWA equation into a polynomial equation with the help of Lagrange multipliers.

In the study of Yager [19] inter-cluster OWA distance is defined as follows.

Definition 2

An OWA distance between the sets A and B is

$$d_{OWA}(A, B) = OWA\{d(x, y)|\forall x \in A, \forall y \in B\} = \sum_{i=1}^{z} w_i d_{(i)} \qquad (20)$$

where w_i are weights of the OWA operator given directly or calculated according to the any distribution function, $z = |A| \cdot |B|$, and $d_{(i)}$ is the ith maximal distance on Cartesian product of $A \times B$.

9   OWA Distance-Based $C \times K$-Nearest Neighbor Algorithm

The well-known nonparametric approach K-nearest neighbor algorithm, puts a new unknown data into the class that shelters highest number of elements within its nearest neighborhood set.

K-NN algorithm gives equal importance to each of the objects in assigning class label to the input vector which is one of the challenges of the K-NN algorithm. Because such an assignment could reduce the accuracy of the algorithm if there is a strong overlapping degree amongst the data vectors. So, K-nearest neighbor algorithm is a sub-optimal procedure. But, it was proven that the error rate for the 1-NN rule is not more than the twice the optimal Bayes error rate, which asymptotically approaches to the optimal rate as K increases, with infinite number of data [4, 6].

As above-mentioned, a point is assigned to a class in which it has the most number of neighbors with K-NN algorithm. The difference of the $C \times K$-nearest neighbor algorithm is that a point is assigned to a class, which K-nearest points set is closest to the classified point. The distance between the point being classified and K-nearest neighbors set is calculated as OWA-distance.

Assume the following representations [10, 16]:

- $X = \{x_1, x_2,..., x_p\}$ is the set of n labeled samples;
- $\{C_1, C_2,..., C_p\}$ is the class labels of the samples;
- x^{new} is the new sample to be classified;
- C_j^K is the K-nearest neighbors to the point x^{new} set in the class C_j.

Then the OWA distance between the point to be classified, x^{new}, and the K-nearest neighbors set C_j^K is calculated as follows:

$$d\left(x^{new}, C_j^K\right) = OWA\left\{d(x^{new}, x)\big| x \in C_j^K\right\} \tag{21}$$

where $d(x^{new}, x)$ is the distance between the points x^{new} and x. In accordance with the above notations, the pseudo-code of the OWA Distance-Based $C \times K$-NN Classification Algorithm is given below:

```
BEGIN
    Start a learning set with separate classes {C₁,C₂,...,Cₚ} ;
    Clear K -nearest neighbors sets Cⱼᴷ, j=1,..,p .
    Input an unclassified input sample xⁿᵉʷ ;
    Set K, 1≤K≤n
    FOR EACH CLASS    Cⱼ∈{C₁,C₂,...,Cₚ}
        Set i = 0 .
        FOR EACH    ∈ Cⱼ
        IF ( < K )THEN
        Assign x  to the K -nearest neighbors set Cⱼᴷ ;
        i = i +1 ;
            ELSE
                Calculate the distance between x and xⁿᵉʷ ;
                IF (x  is closer to  xⁿᵉʷ than any sample in class Cⱼᴷ )
                THEN Delete the farthest sample from the set Cⱼᴷ .
                    Assign x  to set  Cⱼᴷ
            END IF
        END IF
    END FOR
    Calculate the OWA-distance dⱼ between  xⁿᵉʷ and the set Cⱼᴷ ;
    END FOR
        Mark the class with minimum distance dᵣ=minⱼdⱼ;
        Classify x  in the class r of the last minimum found.
END
```

10   Numerical Example

In order to show how OWA based $C \times K$-NN algorithm works in practice, we will consider a numerical example with 21 points, distributed into three classes. Our aim is to classify a new object with coordinates (5.20, 6.45) into a suitable class by using different strategies (see Fig. 4).

First, for each data, its distance from the unclassified (new) sample is calculated as follows:

$$d_{x_i,new} = \sqrt{(x_{i1} - x_{new,1})^2 + (x_{i2} - x_{new,2})^2} \tag{22}$$

Then for each class the smallest K-distances are determined. Suppose that for our example, $K = 3$. So, for each class, nearest 3 neighbors are of concern. Let the 3-nearest neighbor be denoted as X_1^1, X_2^1, X_3^1 for class 1, X_1^2, X_2^2, X_3^2 for class 2,

Fig. 4 Data used in example

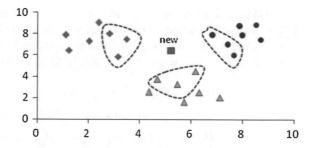

and X_1^3, X_2^3, X_3^3 for class 3. These elements are labeled under the last column of Table 2. In the next step, the distances of the K-nearest neighbors are arranged in a decreasing order for each class and OWA distance is calculated. It is possible to obtain different results which correspond to the linkage strategies given in Sect. 6 by assigning various weights to the corresponding distances. The OWA distance is calculated as follows:

$$OWA(x_{new}, C_j^K) = \sum_{j=1}^c w_j d_{(i),new}^j \tag{23}$$

Table 2 Computation results of the example

ID	x_{i1}	x_{i2}	Class	$d_{x_i,new}$	3-NN_class
1	1.23	6.46	1	3.97	–
2	1.10	7.90	1	4.35	–
3	2.01	7.30	1	3.30	–
4	3.15	5.86	1	2.14	3-NN_1
5	2.38	9.05	1	3.84	–
6	3.48	7.51	1	2.02	3-NN_1
7	2.81	8.04	1	2.87	3-NN_1
8	4.35	2.54	2	4.00	–
9	4.68	3.75	2	2.75	3-NN_2
10	5.46	3.28	2	3.18	3-NN_2
11	5.73	1.61	2	4.86	–
12	6.32	2.53	2	4.08	–
13	6.17	4.53	2	2.15	3-NN_2
14	7.12	2.04	2	4.81	–
15	6.81	7.93	3	2.19	3-NN_3
16	7.41	7.05	3	2.29	3-NN_3
17	7.96	7.93	3	3.13	–
18	7.85	8.81	3	3.55	–
19	7.65	6.04	3	2.48	3-NN_3
20	8.51	8.91	3	4.13	–
21	8.70	7.51	3	3.66	–

where $d_{(i),new}^j$ is the ith furthest distance among K-nearest neighbors in the jth class. For our example, we first take the OWA weights as $w_1 = 1$, $w_2 = 0$, and $w_3 = 0$. This case is compatible with complete-linkage approach and they give the same result. For each class, OWA distances are calculated as follows:

$$OWA(x_{new}, C_1^3) = w_1 d_{(1),new}^1 + w_2 d_{(2),new}^1 + w_3 d_{(3),new}^1$$
$$= 1.0 * 2.87 + 0 * 2.14 + 0 * 2.02 = 2.87$$
$$OWA(x_{new}, C_2^3) = w_1 d_{(1),new}^2 + w_2 d_{(2),new}^2 + w_3 d_{(3),new}^2$$
$$= 1.0 * 3.18 + 0 * 2.75 + 0 * 2.15 = 3.18$$
$$OWA(x_{new}, C_3^3) = w_1 d_{(1),new}^3 + w_2 d_{(2),new}^3 + w_3 d_{(3),new}^3$$
$$= 1.0 * 2.48 + 0 * 2.29 + 0 * 2.19 = 2.48$$

Since min(2.87, 3.18, 2.48) = 2.48, the new sample is assigned to Class 3.

If the OWA weights are changed as $w_1 = 0$, $w_2 = 0$, and $w_3 = 1$, this case is equivalent to the single-linkage approach and OWA distances will be as follows for each class:

$$OWA(x_{new}, C_1^3) = 0 * 2.87 + 0 * 2.14 + 0.1 * 2.02 = 2.02$$
$$OWA(x_{new}, C_2^3) = 0 * 3.18 + 0 * 2.75 + 1.0 * 2.15 = 2.15$$
$$OWA(x_{new}, C_3^3) = 0 * 2.48 + 0 * 2.29 + 1.0 * 2.19 = 2.19$$

Since min(2.02, 2.15, 2.19) = 2.02, the new sample is assigned to Class 1.

For the last case, we get the OWA weights as $w_1 = w_2 = w_3 = 1/3$. This is suitable for the average-linkage approach and OWA distances are calculated as follows for each class:

$$OWA(x_{new}, class1) = \frac{1}{3} * 2.87 + \frac{1}{3} * 2.14 + \frac{1}{3} * 2.02 = 2.34$$
$$OWA(x_{new}, class2) = \frac{1}{3} * 3.18 + \frac{1}{3} * 2.75 + \frac{1}{3} * 2.15 = 2.69$$
$$OWA(x_{new}, class3) = \frac{1}{3} * 2.48 + \frac{1}{3} * 2.29 + \frac{1}{3} * 2.19 = 2.32$$

Among these values, since min(2.34, 3.69, 2.32) = 2.32, the new sample is assigned to Class 3.

If the classical K-nearest neighbor algorithm, for $K = 3$ had been used in the classification process, the new data should have been assigned to Class 1. Because in this approach, the most dominant class among the nearest neighbors is of concern. The nearest three neighbors are x_4 from Class 1, x_6 from Class 1, and x_{13} from Class 3. Since 2 of the nearest neighbor out 3 is from Class 1, the dominating class is Class 1 and the new point assigned to this class.

11   Conclusion

This work is focused on presenting an efficient and flexible computational scheme, OWA distance-based $C \times K$-NN, which is more flexible than the traditional K-NN for classification. A numerical example is solved step by step.

We think that due to its flexibility and adjustability, better results could be obtained if OWA distance based $C \times K$-NN algorithm is applied to any problem instead of any other K-NN based algorithm.

Acknowledgments The authors would like to thank the anonymous reviewers for the constructive discussions and suggestions to improve the quality of this paper. This work is supported by TUBITAK (Scientific and Technological Research Council of Turkey) Grant No. 111T273.

References

1. T.M. Cover, P.E. Hart, Nearest neighbor pattern classification. IEEE Trans. Inf. Theory **13**, 21–27 (1967)
2. T.M. Cover, Estimates by the nearest neighbor rule. IEEE Trans. Inf. Theory **14**, 50–55 (1968)
3. B.V. Dasarathy, Visiting nearest neighbor—a survey of nearest neighbor classification techniques, in *Proceeding International Conference on Cybernetics Society* (1977), pp. 630–636
4. R.O. Duda, P.E. Hart, D.G. Stork, *Pattern Classification* (Wiley, New York, 2001)
5. J. Friedman, Flexible metric nearest neighbor classification, Technical Report 113, Stanford University, Statistics Department (1994)
6. K. Fukunaga, L.D. Hostetler, K-nearest-neighbor Bayes risk estimation. IEEE Trans. Inf. Theory **21**(3), 285–293 (1975)
7. R. Fuller, P. Majlender, On obtaining minimal variability OWA operator weights. Fuzzy Sets Syst. **136**, 203–215 (2003)
8. P. Hart, The condensed nearest neighbor rule. IEEE Trans. Inf. Theory **14**, 515–516 (1968)
9. J. Keller, M.R. Gray, J.A. Givens, A Fuzzy K-nearest neighbor algorithm. IEEE Trans. Sys. Man Cybern. SMC **15**(4), 580–585 (1985)
10. E. Nasibov, C. Kandemir-Cavas, OWA-based linkage method in hierarchical clustering: application on phylogenetic trees. Expert Syst. Appl. **38**, 12684–12690 (2011)
11. R.A. Nasibova, E.N. Nasibov, Linear aggregation with weighted ranking. Autom. Control Comput. Sci. **44**(2), 96–102 (2010)
12. M. Paik, Y. Yang, Combining nearest neighbor classifiers versus cross-validation selection. Stat. Appl. Genet. Mol. Biol. **3**, 1–19 (2004)
13. T.D. Pham, An optimally weighted fuzzy k-NN algorithm, in *ICAPR'05 Proceedings of the Third international conference on Advances in Pattern Recognition,* vol. Part I, (Springer, Berlin, 2005), pp. 239–247
14. B.W. Silverman, *Density Estimation for Statistics and Data Analysis* (Chapman & Hall, UK, 1986)
15. G. Ulutagay, E. Nasibov, A New C×K-nearest neighbor linkage approach to the classification problem, in *Proceedings of the 10th International FLINS Conference*, vol. 1, eds. by C. Kahraman, E.K. Kerre, F.K. Bozbura (World Scientific Proceedings on Computer Engineering and Information Science, Istanbul, Turkey, 2012a) , pp. 471–476, ISBN 978-981-4417-73-0

16. G. Ulutagay, E. Nasibov, OWA aggregation based C×K-nearest neighbor classification algorithm, in *Proceedings of IEEE 6th International Conference Intelligent Systems,* vol. 1, ed. by R.R. Yager, V. Sgurev, M. Hadjiski (IEEE Catalog Number CFP12802-PRT, 2012b), pp. 219–224. ISBN 978-1-4673-2782-4
17. V. Vapnik, *Statistical Learning Theory* (Wiley, New York, 1998)
18. Z. Xu, An overview of methods determining OWA weights. Int. J. Intell. Syst. **20**, 843–865 (2005)
19. R. Yager, On ordered weighted averaging aggregation operators in multicriteria decision making. IEEE Trans. Syst. Man Cybern. **18**, 183–190 (1988)
20. R. Yager, Intelligent control of the hierarchical agglomerative clustering process. IEEE Trans. Syst. Man Cybern. Part B: Cybern. **30**(6), 835–845 (2000)

ARTOD: Autonomous Real Time Objects Detection by a Moving Camera Using Recursive Density Estimation

Pouria Sadeghi-Tehran and Plamen Angelov

Abstract A new approach to autonomously detect moving objects in a video captured by a moving camera is proposed in this chapter. The proposed method is separated in two modules. In the first part, the well-known scale invariant feature transformation (SIFT) and the RANSAC algorithm are used to estimate the camera movement. In the second part, recursive density estimation (RDE) is used to build a model of the background and detect moving objects in a scene. The results are presented for both indoor and outdoor video sequences taken from a UAV for outdoor scenario and handheld camera for indoor experiment.

1 Introduction

Over the last four decades, detection and tracking of moving objects has been studied extensively and it is still a challenging topic in computer vision. There is a wide variety of applications in this area such as, medical imaging, transportation, robots navigation, UAVs, and intelligent vehicle systems. Additionally, it plays an important role in security and surveillance systems. The key feature of any autonomous surveillance system is to enable users to recognise and track activities in the environment where the system is mounted in. In general, there are three possible scenarios for objects detection. In first scenario, a static camera is used to detect and track moving objects. In the second one, moving objects are detected when the camera is also moving. Last configuration is to detect stationary objects by moving cameras. Most of the introduced techniques for detecting moving objects work only for static cameras [1–4]. However, the rising demand for mobile

P. Sadeghi-Tehran · P. Angelov (✉)
School of Computing and Communications, Infolab21, Lancaster University,
City of Lancaster LA1 4WA, UK
e-mail: p.angelov@lancaster.ac.uk

P. Sadeghi-Tehran
e-mail: pouria.sadeghi-tehran@rothamsted.ac.uk

© Springer International Publishing Switzerland 2016
M. Hadjiski et al. (eds.), *Novel Applications of Intelligent Systems*,
Studies in Computational Intelligence 586, DOI 10.1007/978-3-319-14194-7_7

surveillance platforms, such as unmanned aerial vehicles (UAVs) or ground-based vehicles demonstrate the need to detect moving objects by a moving camera(s) more than ever. Since detecting moving objects with moving cameras is much more challenging than a static camera, only a small number of methods have been proposed compared to the case of stationary cameras. In this paper, we tackle the challenge of detecting moving objects with moving cameras by introducing a new approach.

In many computer vision applications background subtraction (BS) is a fundamental part for detection of moving objects. The idea behind this method is to utilise the visual properties of the scene to build an appropriate representation of the expected part of the scene, which is background and separate it from the unexpected part, which is foreground. The information provided by this is used for objects detection and event analysis. One of the most intuitive ways in background subtraction is to compare the colour density of each pixel of an image to the corresponding one in the next frame. If the colour density of two pixels at the same position but in successive frames is the same, they can be classified as a BG. However, if those colours of moving objects are significantly different from those in the background, FG can be classified. In general, to achieve accurate and robust foreground detection, background subtraction should be robust to noise and illumination changes. For instance, car light or swaying tree branches should not be considered and classified as novelty/foreground. In order to increase the robustness of background subtraction technique, several approaches have been introduced which is based on statistical modelling to build background and foreground models [5]. In statistical approaches, data is modelled based on its statistical properties. This information is used later to estimate whether or not a data sample comes from the same distribution. In this method, each pixel in an image is modelled as a random variable in a particular feature space, such as colour along with the probability density function (pdf) in a parametric or non-parametric form. In parametric approaches it is assumed that the data distribution is known and the approach tries to estimate the parameters of this distribution. Any tested data which falls outside the normal parameters of the model is classified as a novelty/foreground. Gaussian Mixture Model (GMM) [6] and Hidden Markov Model (HMM) [7] are among the more achieved methods of this type. On the other hand, non-parametric methods do not require any assumption about the underlying distribution or the statistical properties of data. These approaches are very accurate and flexible; however, due to the need of storing the former pixel value they usually require a huge amount of memory [8]. In addition, in order to model a background a window function is required, which is the same as buffering some frames according to window size and role as a criterion for speculating the background. The correct choice of window size is vital. The choice of too small or too large of window size may deteriorate the accuracy of the system performance. In addition, to distinguish a boundary between foreground and background pixels a threshold is used. A wrong choice of the value of the threshold may result in low performance and distortion of the system in different environment. Nevertheless, the biggest limitation of background subtraction algorithms is that they require the camera to be stationary. If the camera starts moving, the BS algorithm cannot locate the new position of the new pixels relative to the background model. It makes a severe limitation to the range of applications which can be applied using background subtraction.

As mentioned before, a wide variety of applications require a moving camera to identify and track moving objects. Therefore, BS algorithms must be able to accommodate camera movement in order to work in these kinds of applications. A hybrid method based on colour segmentation and motion-based regions is proposed in [9]. Although, this method provides good results, the computational complexity is too high which prevents its real-time application. Murray and Basu [10] used a background compensation method in order to calculate the background motion from the camera pan and tilt angles. However, their method is only restricted to rotation of the camera about the lens centre. On the other hand, another method [11] employs an alternative technique based on the motion estimation of the background by tracking its key points to estimate the parameters of an affine transformation. Optical flow is another commonly used technique to detect moving objects in moving camera scenarios [12, 13]. Dense optical flow fields are used in [14] over multiple frames to estimate the camera motion and segmentation of moving objects. However, inconsistency object boundaries in optical flow methods causes dislocation of the moving objects or split articulate motion of a moving object into more than one object. Other approaches can be used to compensate the camera motion alternatively. One is to build and maintain a background mosaic image. For every new frame, matching pixels in the mosaic image are updated by averaging its value with the value of the corresponding pixel on the warped image. Several feature-based image registration techniques can be used to find the matching keypoints [15–17] between two consecutive images. Once the keypoints are extracted, the motion parameters are estimated using the RANSAC algorithm [18].

2 The Proposed Approach

In this chapter, we tried to overcome the background subtraction limitation by proposing a new technique using the recently introduced recursive density estimation (RDE) approach and camera motion estimation techniques. The whole process is performed in two main phases. In the first phase, the camera movement is compensated and a mosaic is built. In the second phase, the mosaic image is passed to the background subtraction algorithm to identify moving objects (Fig. 1).

2.1 Camera Motion Estimation

Several techniques have been proposed to compensate the camera motion [19–23] to rectify each frame of an image sequences to a coordinate system of a background model. In that case, background subtraction can perform in a normal fashion.

Fig. 1 Schematic diagram of
the introduced approach

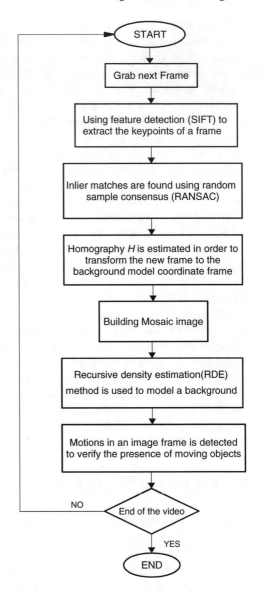

However, some of these approaches perform camera motion estimation and BS of
the entire video in a batch mode which is severely limiting these types of applications [24]. In addition, it is assumed that there are no significant camera orientation changes during the process. On the other hand, our approach can be applied
on real time applications; also, a background subtraction algorithm in our method
can adapt quickly to a rapidly changing background as well as growing its background model when new parts of the scene become visible. In order to compensate
the camera movement, for each consecutive frame a homography H is calculated

which transforms the new frame to the background model's coordinate frame. Then, the homography H and the new frame are fed to the background algorithm. We used a well-known feature based technique known as Scale Invariant Feature Transform (SIFT) [17] to detect feature points in image sequences and to estimate transformation H between frame i and $i + 1$. For each new frame a set of keypoints is extracted and matched with the SIFT keypoints from the previous frame. Then matrix H is estimated using Random Sample Consensus (RANSAC) [18].

2.1.1 Feature Detection Using Scale Invariant Feature Transform (SIFT)

The robustness and distinctiveness of the SIFT approach to detect keypoints in images [17] has been proven in many applications. They are invariant to rotation and image scaling, also, partially invariant to 3D camera viewpoint and illumination changes. In addition, they are well localized in both the frequency and spatial domains, thus, minimizing the disruptions caused by noise, occlusion, and clutter.

The generation of SIFT keypoints is done in four stages. In the first stage, edges of an object or any other highcontrast regions are used to identify interest points in the image. In the second stage, interest points which are stable with respect to invariance to scale and orientation are filtered [13]. These stable interest points are called the SIFT keypoints. In the next stage, each keypoint is assigned one or more orientations which depends on the direction of the gradient present at the sub-pixel level [13]. This process makes each and every keypoint unique and invariant to changes in scale, angular orientation and location. The final stage involves measuring the image gradient around each keypoint thus making the keypoints invariant to shape distortion, noise and illumination [13]. Since these keypoints are unique, a particular keypoint can be matched correctly [17] across many images.

At least 3 keypoints have to be correctly matched, for an object to be matched between two or more frames [17]. Since a very small number of keypoints matches are required, even partially occluded objects can be correctly matched in separate images. While matching keypoints, variation in the location of keypoints in two separate images is also allowed. This makes it possible to match similar objects at different locations in two images and; thus, SIFT keypoints can be used to track a moving object in a video sequence. When the SIFT algorithm is started, it extracts keypoints from the current frame and the previous frame. The keypoints extracted in the first frame are then matched with the corresponding keypoints in other frames of the video sequence.

2.1.2 Homography, RANSAC, and Building a Mosaic Image

Tens or hundreds of keypoints have been extracted from each frame using SIFT; however, when the keypoints are matched there can be few mismatches (Fig. 2). In addition, some of the extracted keypoints in an image frame can be from moving

26 tentative matches

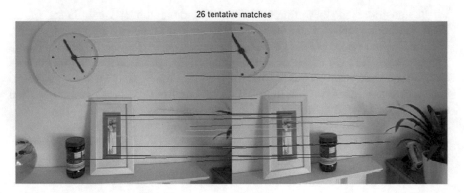

Fig. 2 Finding match points between two frames

objects which do not represent static background scene. In order to refine outliers and remove the incorrect matches between two image frames, RANSAC algorithm is used. RANSAC algorithm tries to find inliers by estimating matrix H and ignoring outliers [18]. It randomly samples corresponding matches and tries to fit the homography H to those data samples. Then, an error is calculated between the rest of data samples and the model. The data points are classified as inlier or outliers based on a threshold T. The process is continued till the number of outliers is sufficiently small [25].

RANSAC method should be able to distinguish how well the estimated H fits all the data samples. In order to do that the symmetric transfer error is used as the distance measure [22].

$$E(z_i, z_i') = d(z_i, H^{-1} z_i')^2 + d(z_i', H z_i)^2 \tag{1}$$

where H is the estimated homography matrix, $d(x_1, x_2)$ is a Euclidean distance between two samples x_1 and x_2, z_i and z_i' arc the coordinates of the corresponding keypoints. If $E(z_i, z_i') < T$, a corresponding pair z_i' and z_i is considered to be an inlier. As shown in Fig. 3 after using RANSAC the number of inlier matches is reduced to 22 out of 28 matches from the beginning while using SIFT only.

Let assume $z = [x_i, y_i]^T$ be the coordinates of a keypoint in the first frame (Fig. 2, left side), and $z' = [x_i', y_i']^T$ be the coordinates of the corresponding keypoint in the next frame (Fig. 2, right side). In that case, the transformation between two frames can be defined by a homography H.

$$z \approx H z' \tag{2}$$

22 (84.62%) Inliner Matches Out of 26

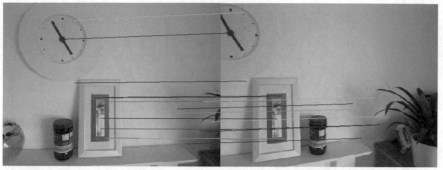

Fig. 3 22 inlier matches is extracted out of 26 matches at the beginning of implementing SIFT algorithm

It should be noted that to estimate the homography H at least 4 match points are needed.

As mentioned before, z_i and z_i' are homogenous vectors and homography H defined up to scale z_i' might not be equal to Hz_i. Hartley and Zisserman solved the equivalent equation $z_i' \times Hz_i = 0$ [25]. Direct Linear Transform algorithm is a well-known method to find homography H [26]. A variant of the direct linear transform described in [22, 25] is used in this work.

If we have four, 2D keypoints between two images, $z_i = [x_i, y_i]^T, z' =_i [x_i', y_i']^T$, and $i \in \{1, \ldots, 4\}$. The aim is to find matrix H in the sense that $z_i' \approx Hz_i$ for all the keypoints [22].

Hz_i can be obtained as follows:

$$Hz_i = \begin{bmatrix} h_{11} & h_{12} & h_{13} \\ h_{21} & h_{22} & h_{23} \\ h_{31} & h_{32} & h_{33} \end{bmatrix} \begin{bmatrix} x_i \\ y_i \\ v_i \end{bmatrix} = \begin{bmatrix} h_{11}x_i + h_{12}y_i + h_{13}v_i \\ h_{21}x_i + h_{22}y_i + h_{23}v_i \\ h_{31}x_i + h_{32}y_i + h_{33}v_i \end{bmatrix} = \begin{bmatrix} \mathbf{h}_1^T z_i \\ \mathbf{h}_2^T z_i \\ \mathbf{h}_3^T z_i \end{bmatrix} \quad (3)$$

In addition, $z_i' \times Hz_i = 0$ can be calculated as:

$$z_i' \times Hz_i = \begin{bmatrix} y_i'\mathbf{h}_3^T z_i - v_i'\mathbf{h}_2^T z_i \\ v_i'\mathbf{h}_1^T z_i - x_i'\mathbf{h}_3^T z_i \\ x_i'\mathbf{h}_2^T z_i - y_i'\mathbf{h}_1^T z_i \end{bmatrix} = 0 \quad (4)$$

$$\begin{bmatrix} y_i'z_i^T h_3 - v_i'z_i^T h_2 \\ v_i'z_i^T h_1 - x_i'z_i^T h_3 \\ x_i'z_i^T h_2 - y_i'z_i^T h_1 \end{bmatrix} = 0 \quad (5)$$

Then, Eq. 5 can be shown as a linear system:

$$\begin{bmatrix} 0 & -v_i'z_i^T & y_i'z_i^T \\ v_i'z_i^T & 0 & -x_i'z_i^T \\ -y_i'z_i^T & x_i'z_i^T & 0 \end{bmatrix} \begin{bmatrix} h_1 \\ h_2 \\ h_3 \end{bmatrix} = \tag{6a}$$

$$\begin{bmatrix} 0 & 0 & 0 & -v_i'x_i & -v_i'y_i & -v_i'v_i & y_i'x_i & y_i'y_i & y_i'v_i \\ v_i'x_i & v_i'y_i & v_i'v_i & 0 & 0 & 0 & -x_i'x_i & -x_i'y_i & -x_i'v_i \\ -y_i'x_i & -y_i'y_i & -y_i'v_i & x_i'x_i & x_i'y_i & x_i'v_i & 0 & 0 & 0 \end{bmatrix} \begin{bmatrix} h_{11} \\ h_{12} \\ h_{13} \\ h_{21} \\ h_{22} \\ h_{23} \\ h_{31} \\ h_{32} \\ h_{32} \end{bmatrix}$$

$$= A_i h = 0$$

$$\tag{6b}$$

In most of the cases, more than four correspondences can be obtained. The position of these points are not absolutely precise; thus, h = 0 is the unique solution to Eq. 6 [25]. The problem is reformulated as the minimisaton of $\|Ah\|$ subject to $\|h\| = 1$. The smallest eigenvector of $A^T A$ is the solution to this optimisation problem [25].

Hartley and Zisserman also mentioned that since direct linear transform minimises algebraic error, it is not invariant to the coordinated image frame for keypoints [22, 25]. Thus, coordinate transformations is used to normalise sets of matching keypoints in a new frame and the background, denoted as T and T'. In addition, the normalised keypoint coordinates can be shown as $\tilde{z}_i = T z_i$ and $\tilde{z}_i' = T' z_i'$.

Homography H for the unnormalised points is as follows:

$$H = T'^{-1} \tilde{H} T \tag{7}$$

where $\tilde{H}$ is the normalised homography which relates to the normalised keypoints $\tilde{z}_i' = \tilde{H} \tilde{x}_i$.

To calculate T and T', each set of coordinates is translated so that the centroid of each set is located at the origin [22]. Then each set is scaled in a way that the mean distance from the points to the origin is $\sqrt{2}$. Normalised matrix T for the coordinates z_i is as follows:

$$T = \begin{bmatrix} \frac{\sqrt{2}}{\mu_d} & 0 & \frac{\sqrt{2}}{\mu_d}\mu_x \\ 0 & \frac{\sqrt{2}}{\mu_d} & -\frac{\sqrt{2}}{\mu_d}\mu_y \\ 0 & 0 & 1 \end{bmatrix} \tag{8}$$

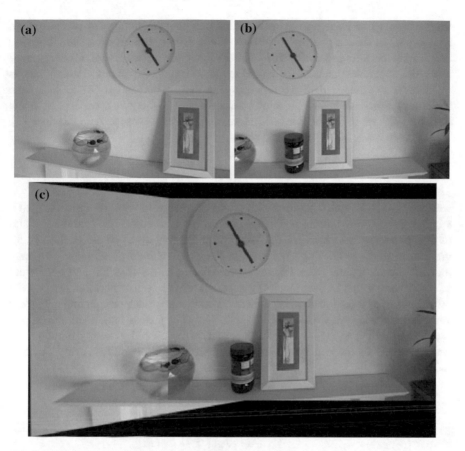

Fig. 4 **a** First image. **b** Second image. **c** Mosaic image

where μ_x and μ_y are the x and y coordinates of the mean, and μ_d is a mean distance from coordinates to the origin. Similarly, normalised matrix T' can be calculated.

Once the homography H is estimated, a new image frame is aligned to the background model coordinates to build a mosaic image (Fig. 4c).

3 Recursive Density Estimation for Novelty/Objects Detection

After motion of a camera is compensated and a new frame is aligned into the coordinate system of the background model, the next step is to apply a reliable technique to model the background to find novelties in the current image frame. The background modelling technique should be fast, computationally efficient, and has a fast response to environment changes without requiring pre-setting any parameters. In addition, when unseen parts of the frame become visible, an algorithm should be able to grow the background model and train the new parts of the frame; also, the algorithm should update pixels of the model to the corresponding pixels in a new image.

In order to achieve that, we use recursive density estimation (RDE) technique [1, 27–29]. The basic technique, kernel density estimation require a window of image frames to be kept in the memory which are not suitable for real-time applications [2].

$$p(s) = \frac{1}{k} \sum_{i=1}^{k} K_\sigma(s - s_i) \tag{9}$$

where K_σ is the kernel function which is also known as a window function with a σ bandwidth. At any initial stage, window function is to buffer few frames. The aim of using window is to estimate the last pixel's distribution to be assigned to background or foreground. The drawback is the small choice of window size may increase the computational efficiency of calculating kernel function and bandwidth of it and may result to deteriorate the accuracy of the system performance; on the other hand, a large value of window size makes the approach computationally slow and non-tractable. Another challenge is to define a proper bandwidth for kernel function. A wide bandwidth results in over-smooth probability density function and a very narrow bandwidth would cause over sensitive probability density function [30].

To address these difficulties the recursive density estimation was introduced [27, 28, 31, 33]:

$$D(s_k) = \frac{1}{1 + \|s_k - \mu_k\|^2 + \Sigma_k - \|\mu_k\|^2} \tag{10}$$

where D denotes density; $\sum$ denotes scalar product of the data samples, s is color pixel, $s = [R, G, B]^T$, $s = [H, S, V]^T$.

The detailed derivation of the expression (10) is provided in [28]. Both, the mean, μ and the scalar product can be updated recursively as follows [28]:

$$\mu_k = \frac{k-1}{k} \mu_{k-1} + \frac{1}{k} s_k \quad \mu_1 = s_1 \tag{11}$$

$$\sum_k = \frac{k-1}{k} \sum_{k-1} + \frac{1}{k} \|s_k\|^2 \quad \sum_1 = \|s_1\|^2 \tag{12}$$

Recursive density estimation technique is non-parametric and it represents the distance from a data sample to all previous samples. The integral of the *pdf* is equal to one while the density according to RDE can have values one and the integral is larger than one [33].

Having the value of the density updated for the current, kth image frame for each pixel one can also calculate and update the mean density, $\bar{D}_k$ as follows [28]:

$$\bar{D}_k = \frac{k-1}{k}\bar{D}_{k-1} + \frac{1}{k}D_k \quad \bar{D}_1 = D_1 \tag{13}$$

One can also recursively update the variance of the density as follows:

$$\left(\sigma_k^D\right)^2 = \frac{k-1}{k}(\sigma_{k-1}^D)^2 + \frac{1}{k}(D_k - \bar{D}_k)^2 \quad (\sigma_1^D)^2 = 1 \tag{14}$$

Based on Eqs. (10–14), foreground pixels can be detected based on a sudden and significantly drop of the density. Once, the foreground pixels are identified (Fig. 5) the density is calculated again. However, this time not in term of time but this time not in terms of time (between pixels in the same position within frames at different time moments as above), but in terms of space—between all foreground pixels of

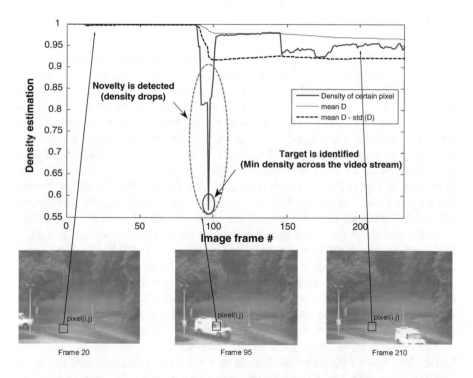

Fig. 5 The evolution of the density of a background pixel throughout the video-stream. The frames for which the value of the density drop below *mean(D) − std(D)* is detected as a foreground pixel [33]

the same frame (note, that their number is significantly smaller than the number of all pixels in a frame) in terms of their horizontal and vertical position. This step of the approach is described in more detail in [31–33].

The advantage of using this method is being free of setting any threshold to find a boundary between foreground and background pixels. Each frame is taken from the camera and after homography H is calculated and a mosaic image is built, it is fed to the RDE algorithm. Since RDE algorithm is threshold-free it is robust to noise and illumination changes. Most importantly, after each image is analysed it is discarded and only information regarding to colour density is stored in the memory which reduces the amount of memory required for processing a frame compared to other methods such as kernel density estimation method. The other advantage is that RDE method does not require training as opposed to other methods like KDE [2] and codebook BS algorithm [34] which usually uses the first 100–200 frames to train the algorithm. The introduced method can be applied from the beginning of the process.

4 Experimental Results

In order to analyse the performance of the introduced approach, two surveillance videos were tested which can be categorised into ground-based and aerial videos. Videos are taken indoors and outdoors to analyse the performance of the approach in both scenarios. In all figures real images and detection results are presented. The results are shown in black and white where black pixels denote background and white pixels denotes foreground objects. In the first experiment, police car was chasing a bike on a motorway. The footage is taken from a police helicopter observing the situation. The frame in Fig. 6a1 shows a bike driving on the motorway. In Fig. 6b1 a police car appears on the screen chasing the bike. After few frames, the police car was getting closer to the bike and chasing it side-by-side (Fig. 6c2). The input video had a resolution of 640 × 360 pixels and 1649 frames.

While the video was running, the keypoints of each frame were extracted by SIFT and match points were detected and refined using RANSAC. The mosaic image was built and passed to RDE algorithm to detect novelties. RDE started detecting moving objects from the beginning without require any pre-training. The detection results for frame 50, 560, and 1230 produced by RDE are presented in Fig. 6a2, b2, and c2. The results show that both moving objects are detected and classified correctly with minimum amount of noise in each frame.

The second experiment was carried out indoor in Infolab21, Lancaster University. Figure 7 shows several frames of a 324 frame sequences with resolution of 480 × 272 pixels. The video sequence was of a person moving along a hall. The

Fig. 6 **a1** Real image, frame #50. **a2** Detected object denoted by *white pixels*, frame #50. **b1** Real image, frame #560. **b2** Detected object denoted by *white pixels*, frame #560. **c1** Real image, frame #1230. **c2** Detected objects denoted by *white pixels*, frame #1230

camera was handheld instead of using tripod and follows the left-to-right motion of the person (Fig. 7).

The illumination changes during the experiment are clearly seen in Fig. 7a1, b1 and c1. The detection results for frames 65, 88, and 220 produced by the RDE method are shown in Fig. 7a2, b2 and c2. Although the illumination changes few times in a short period of time during the experiment, it did not affect the overall performance of the detection algorithm and only few pixels were incorrectly classified as foreground. The RDE algorithm quickly adapted to the changing illumination and background and was able to produce recognisable silhouettes of the moving person in all frames.

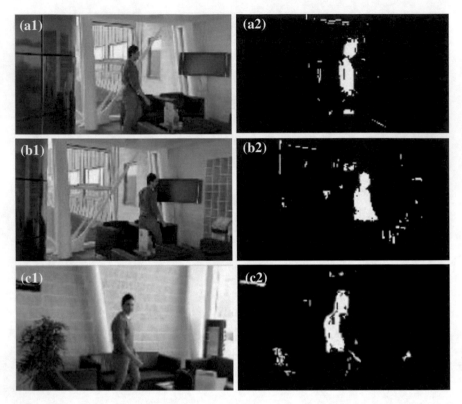

Fig. 7 **a1** Real image, frame #65. **a2** Detected object denoted by *white pixels*, frame #65. **b1** Real image, frame #88. **b2** Detected object denoted by *white pixels*, frame #88. **c1** Real image, frame #220. **c2** Detected object denoted by *white pixels*, frame #220

5 Conclusion

In this work, we extended the ARTOT method [23] which is based on recursive density estimation algorithm (RDE) in order to detect moving objects while a camera is moving. To achieve that, scale-invariant feature transformed and random sample consensus are used to compensate the camera movement. Then, the homography is estimated to transform the new image frame to the background model's coordinate system. The new transformed image passed to the RDE algorithm to detect novelties in the image frame. The proposed approach was tested in indoor and outdoor environments and the results showed satisfactory performance. As opposed to many other background subtraction algorithms such as MoG and KDE, which require several hundred frames for training and model the background before the camera starts moving, RDE can be implemented from the beginning. In addition, in other background subtraction algorithms it is important that each pixel in a new coming frame is compared to the exact position of corresponding pixel in the background model; otherwise it is miss-classified as foreground pixel. However, it is not the case for RDE. Since RDE is threshold-free and the novelty is identified only if it is below $mean(D) - std(D)$ (see Fig. 5), a precise match between a new frame and the background model is not required and some misalignment is allowed. In addition, the introduced approach ARTOD is very robust and can adapt quickly to the environment changes and unseen scene.

References

1. P. Sadeghi-Tehran, P. Angelov, R. Ramezani, *A Fast Approach to Autonomous Detection, Identification, and Tracking of Multiple Objects in Video Streams under Uncertainties* (Springer, Berlin, 2010)
2. A. Elgammal et al., Background and foreground modeling using nonparametric kernel density for visual surveillance. Proc. IEEE **40**, 1151–1163 (2002)
3. B. Leibe et al., Coupled object detection and tracking from static cameras and moving vehicles. IEEE Trans. Pattern Anal. Mach. Intell. **30**(10), 1683–1698 (2008)
4. L. Maddalena, A. Petrosino, Stopped object detection by learning foreground model in videos. IEEE Trans. Neural Netw. Learn. Syst. **24**(5), 723–735 (2013)
5. M. Markou, S. Singh, Novelty detection: a review, part 1: statistical approaches. Sig. Process. **83**, 2481–2497 (2003)
6. C. Stauffer, W. Grimson, Adaptive background mixture models for real-time tracking, in *Computer Vision and Pattern Recognition* (1999)
7. B. Stenger et al., Topology free hidden markov models: application to background modelling, in *Proceedings of IEEE Conference in Computer Vision* (2001)
8. A. Elgammal, D. Harwood, L.S. Davis, Nonparametric background model for background subtraction, in *Proceedings of Conference Computer Vision* (2000)
9. M. Gelgon, P.A. Bouthemy, A region-level motion based graph representation and labeling for tracking a spatial image partition. Pattern Recogn. **33**, 725–740 (2000)
10. D. Murray, A. Basu, Motion tracking with an active camera. IEEE Trans. Pattern Anal. Mach. Intell. **16**(5), 449–459 (1994)
11. S. Araki et al., Real time tracking of multiple moving object contours in a moving camera image sequence. IEICE Trans. Info. Syst. **83**(7), 1583–1591 (2000)
12. I. Celasun et al., 2-D mesh based video object segmentation and tracking with occlusion resolution. Sig. Proc. Image Commun. **16**, 949–962 (2001)
13. A. Wedel et al., Detection and segmentation of independently moving objects from dense scene flow, in *Proceedings of International Conference on Energy Minimization Methods in Computer Vision and Pattern Recognition* (2009), pp. 14–27
14. G. Zhang et al., Moving object extraction with a hand held camera, in *Proceedings of International Conference on Computer Vision* (2006), pp. 1–8
15. H. Bay et al., SURF: speeded up robust features, in *Computer Vision and Image Understanding (CVIU)* (2008), pp. 346–359
16. E. Rublee et al., in *ORB: An Efficient Alternative to SIFT and SURF* (2010)
17. D.G. Lowe, Distinctive image features from scale-invariant keypoints. Int. J. Comput. Vision **60**(2), 91–110 (2004)
18. M.A. Fischler, R.C. Bolles, Random sample consensus: a paradigm for model fitting with applications to image analysis and automated cartography. Comm. ACM **24**(6), 381–395 (1981)
19. D. Farin, W.A. Effelsberg, video-object segmentation using multi-sprite background subtraction, in *Proceedings of IEEE International Conference on Multimedia and Expo* (2004), pp. 343–346

20. E. Hayman, J. Eklundh, Statistical background subtraction for a mobile observer, in *Ninth IEEE International Conference on Computer Vision* (2003), pp. 67–74
21. A. Mittal, D. Huttenlocher, Scene modeling for wide area surveillance and image synthesis, in *IEEE Computer Society* (2000), pp. 160–167
22. E. Tsinko, Background subtraction with a pan/tilt camera, in *The Faculty of Graduate Studies* (The University of British Columbia, 2010)
23. P. Angelov et al., ARTOT: autonomous real-time object detection and tracking by a moving camera, in *Proceedings of 2012 IEEE Conference on Intelligent Systems* (2012)
24. Y. Sugaya, K. Kanatani, Extracting moving objects from a moving camera video sequence, in *10th Symposium on Sensing via Imaging Information* (2004), pp. 279–284
25. R. Hartley, A. Zisserman, in *Multiple view geometry in computer vision*. (Cambridge University Press, 2003)
26. Y. Abdel-Aziz, H. Karara, Direct linear transformation from comparator coordinates into object space coordinates in close-range photogrammetry. Amer. Soc. Photogrammetry, pp. 1–18 (1971)
27. P. Angelov, An approach for fuzzy rule-base adaptation using online clustering. Int. J. Approximate Reasoning, pp. 275–289 (2004)
28. P. Angelov, in *Anomalous System State Identification, Patent Filled, GB1208542.9*, priority date 15 May 2012
29. P. Angelov, R. Ramezani, X. Zhou, Autonomous novelty detection and object tracking in video streams using evolving clustering and Takagi-Sugeno type neuro-fuzzy system, in *IEEE International Joint Conference on Neural Networks* (2008), pp. 1457–1464
30. R. O. Duda, , D.G. Stork, P.E. Hart, in *Pattern Classification*, 2nd edn. (Willey-Interscience, New York, 2000)
31. P. Angelov, P. Sadeghi-Tehran, R. Ramezani, A real-time approach to autonomous novelty detection and object tracking in video stream. Int. J. Intell. Syst. **26**(3), 189–205 (2011)
32. P. Sadeghi-Tehran, Automatic methods for video-streams analysis and self-evolving controllers, in *School of Computing and Communications* (Lancaster University, 2012), p. 195
33. P. Angelov, in *Autonomous Learning Systems: From Data Streams to Knowledge in Real Time* (John Wiley and Sons, New York, 2012). ISBN 978-1-1199-5152-0, 273 pp
34. K. Kim et al., Real-time foreground–background segmentation using codebook model. Real-Time Imaging **11**(3), 172–185 (2005)

Improved Genetic Algorithm for Downlink Carrier Allocation in an OFDMA System

Nader El-Zarif and Mariette Awad

Abstract Different intelligent techniques have been proposed to solve the problem of downlink resource allocation in orthogonal frequency division multiple access (OFDMA)-based networks. These include mathematical optimization, game theory and heuristic algorithms. In an attempt to improve the performance of traditional genetic algorithm (GA) and its heuristics, we propose an improved GA (IGA) that optimizes the search space and GA iterations. Using concepts from ordinal optimization (OO) to determine the stopping criteria and sub-sampling alternatives to generate the initial population, IGA shows faster convergence when applied to downlink carrier allocation in an OFDMA system. IGA workflow also includes a new "swap if better"' mutation operator that replaces the random mutation and a novel fitness function that seeks to maximize the total throughput while minimizing the under-allocation in an attempt to meet the quality of service (QoS) requirements for different types of users. Comparing performance of IGA with different fitness functions published in literature shows improved fairness, comparable throughput and standard deviation. Most importantly IGA is able to better meet the QoS requirements for the different types of users (real time and non real-time) and this, within few milliseconds, making it attractive for real time implementation. Future work plans a parallel implementation of IGA to further improve its computational time.

1 Introduction

In an era of growing mobility and data exchange, the need for bandwidth is increasing exponentially. According to Cisco forecasting, mobile data traffic will be doubling year to year such that between 2009 and 2014 it would have increased by

N. El-Zarif · M. Awad (✉)
American University of Beirut, Beirut, Lebanon
e-mail: mariette.awad@aub.edu.lb; ma162@aub.edu.lb

N. El-Zarif
e-mail: nre06@mail.aub.edu

© Springer International Publishing Switzerland 2016
M. Hadjiski et al. (eds.), *Novel Applications of Intelligent Systems*,
Studies in Computational Intelligence 586, DOI 10.1007/978-3-319-14194-7_8

39 times. Today, smart-phones generate as much traffic as 24 basic-feature phones in view of the introduction of mobile services, such as video streaming, video call, mobile TV, and video conferencing [6].

It is thus becoming nearly impossible to meet all users requirements, without some intelligent optimization techniques. In a typical wireless system, there are multiple mobile stations (MSs) that request services with different quality of service (QoS) requirements. The base station (BS) needs to have a scheduling policy that takes into consideration the different QoS requirements such as data rate, latency and error rate. The packets could also result from heterogeneous types of applications: video streaming, voice over IP (VOIP), web browsing, emails and file transfer protocol (FTP), all of which have different QoS. The scheduler needs to allocate these packets differently so their QoS requirements can be met while taking into account the throughput constraints and when the total number of "feasible" carrier allocation is nearly infinite.

Among the artificial intelligence techniques available, genetic algorithm (GA) was proposed for such optimization problems. GA is a stochastic search technique that is based on the concept of natural selection and survival of the fittest. Traditional GA performance is affected by the settings of the evolutionary operators, the choice and size of the initial population, the fitness function and the maximum allowed number of off-springs; of all which are heuristics that shape the quality of the GA solution.

Motivated to enhance the standard GA workflow, we suggest an improved genetic algorithm (IGA), that proposes a new GA operator, a custom fitness function and guides the selection of the initial population using 4 alternatives of which two depend on ordinal optimization (OO) concept for sub-sampling. The guided initial population helps IGA to converge faster and to a better solution. The random mutation operator is replaced in IGA by a "swap if better" mutation that performs hill climbing to reach the local maxima, while maintaining a "social welfare" between the different solutions. While IGA stopping criteria is defined by OO concepts, IGA uses a novel fitness function that seeks to maximize the total throughput while minimizing the under-allocation i.e., it thrives to maintain performance and fairness. We focus specially on downlink resource allocation for an orthogonal frequency division multiple access (OFDMA) system because it is used in the latest wireless standards such as WIMAX and LTE.

Comparing performance of IGA with different fitness functions published in literature shows improved fairness, comparable throughput and standard deviation. Most importantly, IGA is able to meet the QoS requirements for heterogeneous types of users within few milliseconds.

The rest of this paper is organized such that the literature review is presented in Sect. 2 and the problem formulation is detailed in Sect. 3. OO and IGA concepts are explained in Sect. 4, while IGA workflow is validated though simulations in Sect. 5. Finally, in Sect. 6, research conclusions are highlighted.

2 Literature Review

In the following literature, we first provide an overview of the previous research that studied scheduling in wireless systems. Then we present the most relevant work for downlink resource allocation in OFDMA systems using GA.

2.1 Scheduling in Wireless Systems

Concerning scheduling in wireless systems, [20] is the most relevant work addressing the scheduling constraints. Authors in [20] reported that schedulers must be consistent with the system at hand—which in their case was WiMAX, meet fairness among users, satisfy the QoS requirements of each service class, maximize system throughput, minimize power consumption, and be as simple as possible. As contradictory and somewhat redundant as these constraints seem to be, these goals must be met to ensure scalability.

The first issue is how to achieve fairness among users. Researchers in [26] proposed a two-stage fair and effective queueing (FEQ) algorithm: the first stage used weighted round robin to provide minimum rate required (MRR) for each user, while the second phase utilized the "earliest deadline first" algorithm for the remaining bandwidth to minimize the packet dropping rate. Despite the fact that fairness is a very important issue, the algorithm didn't take the channel state information (CSI) into consideration. Indeed, a user with a bad channel condition would consume a larger proportion of the available bandwidth to achieve its MRR, which decreases the throughput of the system.

The second issue is how to mitigate the channel effect while meeting the QoS requirements for real-time polling service (rtPS) users. Prasath et al. [19] proposed a solution based on introducing a parameter α that is used to control the bandwidth division between real-time and non-real-time traffic. As the distance between BS and MS increases, the delay increases, and hence the QoS of real-time traffic decreases; as a result, α varies to give higher priority to real-time traffic to increase its QoS without causing congestion for the non-real-time polling service (nrtPS) buffer.

Another approach to maximizing throughput and maintaining the QoS was performed in [1]. The authors formulated a scheduling algorithm that consists of combining a temporary removal scheduler with a modified maximum signal to interference ratio (mmSIR). Their result showed that mmSIR had a similar performance as mSIR in terms of overall system throughput and number of served subscriber stations (SS) per frame, but showed an improvement in terms of mean sojourn time which helps to meet the QoS requirement of rtPS users.

To reduce the frame occupation ratio, authors in [12] proposed an algorithm based on channel learning. Although it was designed for WiMAX, the algorithm didn't take the queueing delay into consideration. To meet the QoS requirements of service classes while maintaining the system throughput, and to increase the overall transmission control protocol throughput while achieving fairness among users, the authors in [28] proposed a joint optimization between link adaptation and automatic repeat request.

The issue of ensuring fairness while satisfying the QoS requirement of each service class, was tackled by Jayaparvathy et al. [11] from a different angle. The authors suggested an approach based on a dynamic weight adjustment scheduling algorithm derived from Nash equilibrium. Every SS supports four classes of traffic: unsolicited grant service, rtPS, nrtPS and best effort (BE). As each class has its own queue and its own weight, users compete for the provided bandwidth. The weights depend on the QoS specification and congestion. As nrtPS traffic increases, its weight rises, and hence, the QoS of rtPS and BE traffic drops accordingly. Although the link budget was used in the simulation, the modulation order was not taken into consideration. And this approach failed short of resolving the issue of maximizing throughput and minimizing power consumption.

Other game-theoretical based approaches were introduced by [8, 17, 18] to solve the issue of throughput maximization and resource allocation among users. In [8], scheduling is performed in two steps: inter-class and intra-class scheduling. In the former class, the users sort their packets in decreasing order of utility, while in the latter, the scheduler divides the available resources among the users by using the principle of game theory. Simulation results showed that the Nash solution provides a good fairness among the users. In [18, 27] Nash bargaining solution is used where the BS enforces cooperation among the users. The BS chooses the users whose rate maximizes the summation of utilities, where the utility function is linear [18] or logarithmic [27] to ensure proportional fairness among users versus relative throughput maximization. The authors in [18] showed that the Nash bargaining solution (NBS) ensures synchronized fairness and high sum-rate. However, as the number of connections increases, the sum-rate decreases to nearly approximate the max-min fairness solution, and the fairness variance decays to zero. In [27], it was shown that NBS is just another version of proportional fairness.

A different game theoretical approach in [17] is based on a non-cooperative game which proposes to solve the scheduling problem by correlating call admission and bandwidth allocation among users. This correlation makes sure that the QoS doesn't drop below a certain predefined threshold. For real-time applications, the QoS is the delay, whereas for non-real-time applications the QoS is the throughput.

2.2 Resource Allocation Using GA

Concerning scheduling in GA, we present in what follows the most relevant work for downlink resource allocation in OFDMA systems using GA.

Generally speaking, the resource allocation objective is to dynamically distribute subcarriers, bit and power according to the instantaneous channel condition for every user. Song et al. [21] performed joint optimization between the scheduling and link adaptation using GA, disregarding the delay and QoS requirements of each service class.

To maximize system throughput, [24] proposed the waterfilling algorithm as a fitness function for GA. A random initial population was created, then a good individual was added; which helped GA converge faster to a suboptimal solution. However, the waterfilling algorithm did not take into account real-time users. The proposed objective function maximized only the total transmitted power as shown below:

$$\text{Fitness 1} = \underset{b_{k,n}}{\text{argmax}}\left(\sum_{n=1}^{N_C}\sum_{k=1}^{K}\left(\frac{f(b_{k,n})}{\alpha_{k,n}^2}\right)\right) \tag{1}$$

where $b_{k,n}$ is the number of bits allocated to user k using subcarrier n, and $\alpha_{k,n}$ the channel gain of user k using subcarrier n.

$$f(b_{k,n}) = \frac{N_0}{3}\left[Q^{-1}\left(\frac{BER_n}{4}\right)\right]^2(2^{b_{k,n}} - 1) \tag{2}$$

where BER_n is the bit error rate for subcarrier n, Q^{-1} is the inverse Q function, and N_0 is the noise power spectral density.

To achieve fairness among real-time and non-real-time users, authors in [16] proposed a sigmoid-like function and assumed a model of K users of which K' are real time users. Their objective was to maximize the utility as shown below:

$$\text{Fitness 2} = \max\left(\sum_{k=1}^{k=K}U_k(r_k)\right) \tag{3}$$

where r_k is the data rate allocated to user k. The utility is given by:

$$U_{real} = \begin{cases} 0 & r \leq l_1 \\ \sin^k\left(\frac{\pi}{2}\times\frac{r-l_1}{l_2-l_1}\right) & l_1 < r \leq l_2 \\ 1 & r > l_2 \end{cases} \tag{4}$$

$$U_{non-real} = \begin{cases} \log(1+10^{-6r}) & r \leq l_3 \\ 1 & r > l_3 \end{cases} \tag{5}$$

where they considered l_1 = 250 Kbps, l_2 = 5 Mbps, l_3 = 9 Mbps.

Teng et al. [22] proposed an improved complexity reduced genetic algorithm (CRGA) to achieve fairness among real-time traffic users and non-real-time traffic users while maximizing the system utility. In CRGA, the chromosome data is not binary anymore (instead a vector is used to note user index for each carrier) which reduces the chromosome size making the crossover operation much simpler. Because of that, CRGA converged much faster than traditional GA. The algorithm allocated more carriers to real-time users when the channel condition was bad, and increased the data rate of non-real-time users when the channel was good. However, [22] did not compensate for users who receive low data rate during a scheduling period. Teng et al. [22] proposed a model of K users of which K' are real time users and the following proposed fitness function to be maximized:

$$\text{Fitness } 3 = \underset{r_k}{\text{argmax}}(A(r_k) + \lambda \times B(r_k)) \tag{6}$$

$$A(r_k) = \sum_{k=1}^{k=K'} U_k(r_k) \tag{7}$$

$$B(r_k) = \sum_{k=K'+1}^{k=K} U_k(r_k) \tag{8}$$

where λ is a parameter that is used to vary the weight of real-time users, and r_k the total number of bits allocated to user k. Note that if the overall channel is bad, λ should change to give more weight for real time users. λ is defined by:

$$\lambda = \frac{A(r_k)}{B(r_k)} \tag{9}$$

And the utility is given by:

$$U_{real} = \begin{cases} 0 & r \leq 20 \\ \dfrac{1}{1 + e^{10 - 0.08r}} & 20 < r < 200 \\ 1 & r \geq 200 \end{cases} \tag{10}$$

$$U_{non-real} = \begin{cases} 0.5 \log_{10}(1 + 10^{-2r}) & r < 638 \\ 1 & r \geq 638 \end{cases} \tag{11}$$

To perform scheduling in OFDMA system, authors in [7] modify GA parameters in order to increase the system throughput and fairness among users such as the initial population, mutation operator and the stopping criteria. Simulations showed the modified GA operators performed better than the traditional over different fitness functions.

Another way to achieve fairness among users while maximizing the utility is proposed by the authors in [29] who conceived a joint optimization based on GA. The stochastic approximation was used to control the parameters in scheduling, and GA was utilized to improve the carrier allocation among users. Simulations showed that their algorithm outperformed maximum-largest weighted delay first (M_LWDF) in terms of average delay and packet loss rate. Even though the delay was reduced, which makes the algorithm suitable for real-time services, the authors didn't show how the data rate improved.

To meet the QoS requirement of IEEE 802.16 users, Chiu et al. [4] suggested GA with subscriber station (SS) grouping resource-allocation (GGRA). Since most of 802.16 traffic is delay-sensitive, [4] used the notation of residual life-time to increase the priority of users that did not receive allocation during a frame period, so that the QoS of all user classes are met. The algorithm first aggregates highly correlated SS together by means of virtual MIMO systems, and then assigns each SS from the same group to different slots to prevent high mutual interference and minimize the number of genes in GA, which in turn performs allocation based on the increasing order of residual life-time. Their result showed that GGRA outperformed both M_LWDF and efficient and fair scheduling (EFS) in terms of system throughput, ratio of unsatisfied hypertext transfer protocol users and FTP throughput. Even though the GGRA run-time was higher than that of EFS, it allocated carriers within the 5-ms frame duration requirement.

From a GA performance improvement, we report on few works that are most related to IGA. Authors in [14] developed an adaptive real code GA which first specified the key parameters in GA such as: crossover probability, mutation probability, selection size and population size; then classified the parameters into important ones that affect the performance of GA, and unimportant ones that have very little effect on GA. It also classified the important parameters into sensitive (have different effect on GA at different stages) and robust (have similar effect on GA at different stages). By identifying the sensitive parameters, the algorithm would modify them throughout the process as need be.

To the best of our knowledge, authors in [15, 23] were the only one to combine OO and GA to solve mixed integer programming (MIP) and flow shop scheduling respectively. The MIP and flow shop scheduling are non-deterministic polynomial time class problems. Thus OO was used to determine the number of iterations for GA given the required performance of the solution and the confidence level. The number of generations in GA was proportional to the number of iterations provided by OO and inversely proportional to the chromosome size. After determining the size of the initial population and the number of generations, the authors selected a suitable mutation operator. In [15], the mutations were achieved by using a triangular distribution since the data is of integer and not binary type; then, in the final stage the MIP problem was transformed to a linear programming (LP) problem which can be easily solved. In [23], two stages of mutations were used to improve GA results. Simulations were done in production planning, scheduling in batches and in flow shop to validate the proposed models.

Finally, to mitigate the long convergence time of GA, the authors in [3] proposed an algorithm that combines Karush-Kuhn-Tucke (KKT) conditions [2] with GA to perform resource allocation in wireless mesh networks. The algorithm proposed in [3] simply compares KKT-driven approach with GA after 10 iterations, and then chooses the best one. The performance of the combined KKT-GA was proven to be lower-bounded by KKT and upper-bounded with GA with a relatively low order complexity.

3 Problem Formulation

IGA objectives are: total throughput and fairness. Though contradictory as goals, IGA includes them in the same objective function as shown later in this section. But before proceeding further with the problem formulation, we introduce the nomenclature adopted. Let:

- K, N and P represent the total number of users, total number of carriers and maximum available transmitted power at the base station respectively.
- $p_{k,n,t}$ be the transmit power allocated to user k on subcarrier n and scheduling frame t.
- $c_{k,n,t}$ be the allocation of user k on subcarrier n and scheduling frame t.
- $\gamma_{k,n,t}$ be the channel quality of user k on subcarrier n and scheduling frame t.
- $M_{k,n,t}$ be a parameter that is proportional to the number of bits per subcarrier n for user k at scheduling frame t.
- $\rho(.)$ be the data rate which can be achieved with the transmit power $p_{k,n,t}$.
- $H_{k,n,t}$ be the channel gain of subcarrier n if user k is using subcarrier n at scheduling frame t.
- N be the noise power spectral density.
- $SNR_{k,n,t}$ be the received signal to noise ratio at user k if it is using carrier n at a scheduling frame t.
- $N_{k,t}$ be the desired data rate to user k at scheduling frame t.
- $T_{k,t}$ be the maximum time for the user to get serviced after a scheduling frame t.
- T_f be the frame duration.
- $X_{k,t}$ be the maximum number of frames for the user to get serviced after a scheduling frame t.
- $D_{k,t}$ be the demand of user k at scheduling frame t.
- $U_{k,t}$ be utility of user k at scheduling frame t.
- $R_{k,t}$ be the dissatisfaction with the service for user k at a scheduling frame t.

The objective is to find the carrier allocation $c_{k,n,t}$ that minimizes the total sum of user under-allocation or the dissatisfaction level with the service over large number of iterations.

$$\underset{c_{k,n}}{\arg\min} \frac{\Sigma_{\text{all runs}}\Sigma_{\text{all users}} \left(\text{Users under allocation}\right)}{\# \text{ of runs} \times \Sigma_{\text{all users}} \text{ requested rate}} \tag{12}$$

To meet the objective in Eq. 12, we propose the following fitness function:

$$\text{Proposed Fitness} = \underset{c_{k,n,t}}{\arg\max} \sum_{k=1}^{K} \left(U_{k,t}(c_{k,n,t}) - R_{k,t}\right) \tag{13}$$

subject to the following constraints:

1. For a given frame, a carrier can only be allocated to one user.

$$\sum_{k=1}^{K} c_{k,n,t} = 1 \quad \forall n \in [1,N], \quad \forall t \tag{14}$$

2. For all frames, the number of carriers allocated to all users cannot exceed the total number of carriers.

$$\sum_{k=1}^{K} \sum_{n=1}^{N} c_{k,n,t} \leq N \quad \forall t \tag{15}$$

3. The transmitted power cannot exceed the total base station power.

$$\sum_{k=1}^{K} \sum_{n=1}^{N} p_{k,n,t} \leq P \quad \forall t \tag{16}$$

From which we infer that:

$$X_{k,t} = \left\lfloor \frac{T_{k,t}}{T_f} \right\rfloor \tag{17}$$

As $X_{k,t}$ decreases, the demand of the user is higher because it needs the allocation before a degradation of the quality of service occurs. A_1 and A_2 are parameters that can be tuned in order to modify the priority of real-time applications, while A_3 is a parameter that can be tuned to increase the penalty of under-allocation. For non-real-time users $X_{k,t}$ is ∞. +1 is added to $X_{k,t}$ because we don't want $D_{k,t} \rightarrow \infty$ as $X_{k,t} \rightarrow 0$; if that happens, user k will be given the priority forever which will negatively affect the scheduler. $U_{k,t}$ should have a similar form as $D_{k,t}$.

To achieve fairness, we propose adding a parameter to increase the utility of those that did not receive enough carriers to meet their requested QoS in the last frame. If the BS knows the requested data rate of the user, and due to poor channel conditions or congestion, the MS would not assign enough carriers to achieve its QoS; then the BS will store the difference and include it in $R_{k,t-1}$ as the residue term representing dissatisfaction with the service for the next scheduling frame.

$$U_{k,t} = \left[M_{k,t} \left(A_1 + \frac{A_2}{X_{k,t} + 1} \right) \right] - A_3 \times R_{k,t-1} \tag{18}$$

If the user is in a bad channel condition, s/he might not receive enough carriers to meet the requested QoS requirements. By decreasing $R_{k,t-1}$ from the utility, we are forcing the scheduler to give more priority for users that did not receive enough downlink carriers in the previous frames. $M_{k,t}$ is a parameter that is proportional to the data rate of each user.

$$M_{k,t} = \sum_{n=1}^{n=N} C_{k,n,t} M_{k,n,t} \qquad (19)$$

where

$$\begin{cases} C_{k,n,t} = 1 & \text{if user } k \text{ is allocated carrier } n \\ C_{k,n,t} = 0 & \text{Otherwise} \end{cases} \qquad (20)$$

and

$$M_{k,n,t} = \begin{cases} 1 & 4-QAM & R=\frac{1}{3} & SNR<4 \\ 1.5 & 4-QAM & R=\frac{4}{3} & 4<SNR<6 \\ 2 & 16-QAM & R=\frac{1}{2} & 6<SNR<10 \\ 3 & 16-QAM & R=\frac{3}{4} & 10<SNR<14 \\ 4 & 64-QAM & R=\frac{2}{3} & 14<SNR<16 \\ 4.5 & 64-QAM & R=\frac{3}{4} & SNR>16 \end{cases} \qquad (21)$$

Note that the value of $M_{k,n,t}$ depends on the channel condition of user k using carrier n at a scheduling time t. To know the modulation order of the receiver, the received signal power or the received signal to noise ratio is found using:

$$SNR_{k,n,t} = p_{k,n,t} \times \frac{|H_{k,n,t}|^2}{N_{k,n,t}} \qquad (22)$$

A suitable formulation of the residue is one that stores the difference between the demand and the utility only if the demand is greater than the utility. In other words, we want to compensate for the users who did not receive enough scheduling, rather than punish those that did. Thus we propose:

$$R_{k,t} = \begin{cases} D_{k,t} - U_{k,t} & D_{k,t} - U_{k,t} > 0 \\ 0 & \text{Otherwise} \end{cases} \qquad (23)$$

4   Improved Genetic Algorithm

Besides its innovative fitness function, IGA seeks to reduce the heuristics of traditional GA. OO concept is used in order to specify the stopping criteria. Three initial selection algorithms, that depend on sampling from the search space or on sampling from a reduced search space, are proposed to help IGA converge quickly to the desired solution. That is because if the initial population is good enough, IGA should be able to reach the desired solution in fewer generations. Finally, the mutation operator is replaced with a "swap if better" one.

4.1   Ordinal Optimization Concept

OO main idea is that, instead of searching for the best result, a good enough solution with high probability is sought. Its concepts are explained in details in [9, 13]. In a nutshell, the procedures of OO horse race (HR) selection rule are as follows:

1. Sample N designs from the search space uniformly and randomly. It is assumed that the sampled distribution will have the same statistical characteristics of the search space.
2. Use a crude model to estimate the performance of these N designs.
3. Specify the size of the good enough set and the alignment level K (how many good designs are needed to be selected).
4. Identify the most appropriate order performance curve (OPC) class of the problem based on the plots in Fig. 1 and estimate the noise level of the crude model (low, medium or high).
5. Calculate the size of the selected set S according to Eq. 24 and Table 1. It is to be noted that this table is only applicable for a value of N around 1000 [10].
6. Select the observed top S designs in N.

According to [13], the size of the selected subset can be found by the use of statistical methods and regression:

$$S = Z(G, K) = e^{Z_1} K^{Z_2} G^{Z_3} + Z_4 \qquad (24)$$

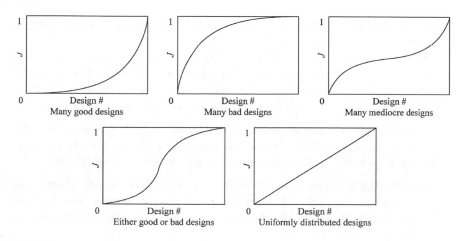

Fig. 1 Five types of the OPC classes [10]

Table 1 Five types of the OPC classes [10]

Noise	∞	U(−0.5, 0.5)				
OPC class	B-pick	Flat	U-shape	Neutral	Bell	Steep
Z_1	7.8189	8.1378	8.1200	7.9000	8.1998	7.7998
Z_2	0.6877	0.8974	1.0044	1.0144	1.9164	1.5099
Z_3	−0.9550	−1.2058	−1.3695	−1.3995	−2.0250	−2.0719
Z_4	0.00	6.00	9.00	7.00	10.00	10.00
Noise	∞	U(−1, 1)				
OPC class	B-pick	Flat	U-shape	Neutral	Bell	Steep
Z_1	7.8189	8.4299	7.9399	8.0200	8.5988	7.5966
Z_2	0.6877	0.7844	0.8989	0.9554	1.4089	1.9801
Z_3	−0.9550	−1.1795	−1.2358	−1.3167	−1.6789	−1.8884
Z_4	0.00	2.00	7.00	10.00	9.00	10.00
Noise	∞	U(−2.5, 2.5)				
OPC class	B-pick	Flat	U-shape	Neutral	Bell	Steep
Z_1	7.8189	8.5200	8.2232	8.4832	8.8697	8.2995
Z_2	0.6877	0.8944	0.9426	1.0207	1.1489	1.3777
Z_3	−0.9550	−1.2286	−1.2677	−1.3761	−1.4734	−1.4986
Z_4	0.00	5.00	6.00	6.00	7.00	8.00

Z_1, Z_2, Z_3 and Z_4 are constants of regression that depend on the OPC class and noise level. Their values can be found according to Table 1. If the top S designs from the N designs are selected, the theory of OO ensures that at least K good designs will be obtained from S with a probability of at least 95 %.

4.2 Improved Genetic Algorithm Workflow

The structure of IGA is shown in Fig. 2 while the chromosome structure is shown in Fig. 3.

The chromosome length is equal to the number of carriers. The index value is the carrier ID while the value pointed by the index is the user ID. As long as the user ID is $\in [1, N_{USERS}]$, the chromosome will always generate a solution inside the search space. In what follows, we detail the six different steps of our proposed IGA.

1. Initial Selection: To sample the initial population from the large search space, we propose three different algorithms: modified random start (MRS), Hammersley approximation (HA) and the uniform sampling inspired method (USIM).

   a. MRS: The MRS pseudo-code is shown in Fig. 4. It consists of allocating carriers to the users who didn't receive enough carriers to meet their requested QoS. Once this objective is met, all remaining carriers are randomly allocated to all users.

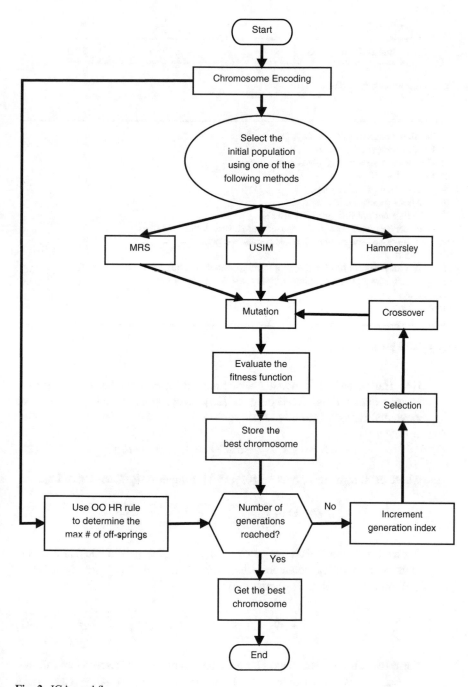

Fig. 2 IGA workflow

Carrier ID	1	2	3	4	⋯	1024
User ID	1	5	6	8	⋯	5

Fig. 3 Chromosome structure

```
1: Get the number of carriers (N_carriers).
2: Get the number of users (N_users).
3: Get the requested throughput for each user.
4: for n = 1 → N_carriers do
5:     Select a user x at random.
6:     Allocate carrier n to user x
7:     Update current throughput of user x
8:     if user x current throughput > user x requested throughput then
9:         Remove user x from the random selection pool.
10:    end if
11:    if If all the users receive throughput more than they requested then
12:        Allocate all the remaining carriers to the users at random.
13:    end if
14: end for
```

Fig. 4 MRS pseudo-code

b. HA: Halton and Hammersley techniques are useful methods to uniformly sample data points. According to [25], every positive integer k can be expanded using a prime base p:

$$k = a_0 + a_1 p + a_2 p^2 + a_3 p^3 + \cdots + a_r p^r \tag{25}$$

where each a_i is an integer in $[0, p-1]$. Function $\Phi_p(k)$ is defined as:

$$\Phi_p(k) = \frac{a_0}{p} + \frac{a_1}{p^2} + \frac{a_2}{p^3} + \cdots + \frac{a_r}{p^{r+1}} \tag{26}$$

For a d dimensional data one can have $p_1, p_2, \ldots, p_{d-1}$, then one can compute their corresponding sequence $\Phi_{p_1}(k), \Phi_{p_2}(k), \ldots \Phi_{p_{d-1}}(k)$, finally a set of n Hammersley points is obtained by:

$$\left(\frac{k}{n}, \Phi_{p_1}(k), \Phi_{p_2}(k), \ldots \Phi_{p_{d-1}}(k) \right) \tag{27}$$

For a two dimensional data, Hammersley points are represented as follows: $\left(\frac{k}{n}, \Phi_{p_1}(k) \right)$. We propose to make the sequence of $\Phi_{p_1}(k)$ indicate the carrier allocation of chromosome x. The set of Hammersley points obtained in the previous section are uniformly distributed, but they are bounded between

0 and 1. By multiplying $\Phi_{p_1}(k)$ by k, one can obtain uniformly sampled points that $\in [0, k]$. Since the resultant value is not an integer, a floor function is needed to make the value represent the user indexes. Therefore, HA results are:

$$\begin{cases} \text{User Index} = \lfloor \Phi_{p_x}(k) \times k \rfloor \\ \text{Carrier Index} = k \end{cases} \tag{28}$$

Note that the following result is for a given prime base p_x, and it represents a single chromosome. To generate the initial population, the process is repeated $N_{chromosomes}$ times for some different base values. Note that $N_{chromosomes}$ is the number of chromosomes in the initial population.

c. USIM: Assuming the size of the initial population is selected to be 10. For 10 users and 1024 carriers the size of the entire search space is 10^{1024} which is very large and far beyond the range of OO. In order to sample from such a large space, a hierarchical approach is used. First, a smaller model is used which consists of 8 users and 12 carriers. The reduced search space size in this case is $8^{12} = 68719476736$, which is within the range of OO. Then, the reduced search space is expanded to the size of the entire search space in order to get the initial population. To uniformly sample from the reduced space (8 users and 12 carriers), the algorithm presented in Fig. 5 is adopted. USIM pseudo-code is detailed in Fig. 6. The reduced search space generates 1000 samples. USIM first selects four users at random. Then it chooses the samples that produces the highest throughput for those users. If the user got his/her requested data rate, s/he would be removed from the selection pool and won't be allocated any additional carriers. The process repeats until all the carriers are allocated. Note that USIM is used to generate one chromosome. The process is repeated $N_{chromosomes}$ times in order to create the initial population. This can be a time consuming task, and might not be suitable for real time scheduling implementations.

```
1:  Select the number of samples N_S {The number should be around 1000}
2:  Set the value of K and N {K = 12 and N = 8}
3:  Step Size ← K^N / N_Samples
4:  Initial Step ← Random Number that is < 0.1 × Step Size
5:  i ← Initial Step
6:  Generate an empty array V of length N_S
7:  for j = 1 → N_S do
8:      Transform index i to base K
9:      V[j] ← i
10:     i ← i + ε {ε is a random number << Step Size}
11:     j ← j + 1
12: end for
```

Fig. 5 Uniform sampling pseudo-code

1: Get the requested throughput for each user.
2: Create dynamic vector U of length K, containing all the numbers from 1 to K in random order. {K = 12 in this simulation}
3: Generate the array V from figure 5 {V contains $N_S = 1000$ uniform samples}.
4: Create an empty array W of length $N = 8$.
5: $N_{S-Ch} \leftarrow \frac{N_{Chromosomes}}{N}$
6: **for** $i = 1 \rightarrow N_{S-Ch}$ **do**
7:    Randomly shuffle U and store its first N values in W
8:    **for** $j = 1 \rightarrow N_S$ **do**
9:      $A[j] \leftarrow V[j]$ {A[j] contains 12 digits. The value every one of them is a positive integer $\in$ [0,7]}
10:      Thr = 0
11:      **for** $k = 1 \rightarrow K$ **do**
12:        Thr $\leftarrow$ Thr + Channel_Cond[i$\times$K][W[k]]
13:        k $\leftarrow$ k+1
14:      **end for**
15:      After selecting the subcarrier that maximizes the throughput: $A[j] \leftarrow W$ {A[j] now contains K digits $d_0, d_1, \ldots d_{K-1}$ valued as W[0], W[1] $\ldots$ W[K-1] respectively. The array A[j] indicates that carrier 0 is allocated to user d_0 and carrier 1 is allocated to user d_1 and so on and so forth}.
16:      j $\leftarrow$ j+1
17:    **end for**
18:    Store A[j] in chromosome n.
19:    Update the received throughput for every user.
20:    **if** If all the users receive throughput more than they requested **then**
21:      Reset the value of U {To contain all intergers from 1 to K}
22:    **else**
23:      Remove the entries from U which corresponds to users who received throughput more than they requested
24:    **end if**
25:    **while** length(U) $\leq K$ **do**
26:      Select at random a value from U and concatenate it to U. U will now contains duplicate values which increase the priority of the selected under-allocated user
27:    **end while**
28:    i $\leftarrow$ i+1
29: **end for**

Fig. 6 USIM pseudo-code

2. Mutation: The "swap if better" mutation operator seeks to improve the "social welfare" among users. If both users are over-allocated or under-allocated, carrier swapping will occur if it is beneficial. If one user is over-allocated and the other is under-allocated, the carrier is taken from the over-allocated user and given to the under-allocated one, while priority of real time users is preserved due to our proposed fitness function. The swap pseudo-code is shown in Fig. 7.
3. Crossover: A single point crossover is used, where the crossover point is selected at random for every generation and every chromosome.
4. Selection: The selection criteria is roulette wheel selection based on the highest fitness. All chromosomes are subject to replacement. Initially the chromosome

```
 1: Select two carriers at random (w,x) and get their corresponding users ID (W,X) and channel
    conditions (C[w][W],C[x][X]).
 2: Get the current throughput of users W and X.
 3: Get the requested throughput for each user.
 4: if user W is under-allocated and user X is over-allocated then
 5:     Allocate carrier x to user W
 6: end if
 7: if user X is under-allocated and user W is over-allocated then
 8:     Allocate carrier w to user X
 9: end if
10: if user W and user X are under-allocated or user W and user X are over-allocated then
11:     Y ← C[w][W] + C[x][X]
12:     Z ← C[w][X] + C[x][W]
13:     if Y > Z then
14:         Allocate carrier w to user X and carrier x to user W
15:     end if
16: end if
```

Fig. 7 "Swap if Better" pseudo-code

with the highest fitness is stored, but it is always replaced by the most fit chromosome in subsequent offsprings.

5. Stopping Criteria: The stopping criterion is the same as presented in [15]. If the size of the selected set S is known, it can determine the size of the initial population $S_{initial}$ or the stopping criteria $N_{generations}$ by the following equation:

$$S = S_{initial} \times N_{chromosomes} \qquad (29)$$

6. Repetition: The process repeats until the number iterations S that is determined by OO is reached.

5   Simulation

5.1   System Assumptions and OFDMA Parameters

We consider a single cell downlink OFDMA system with 10 users and 1024 subcarriers. The users are divided into two classes: real-time and non-real-time users. The BS must perform scheduling and carrier allocation to the users by satisfying the following design specifications:

- Maximize the system throughout.
- Meet the QoS for all types of services.
- Maintain fairness among users.
- Minimize power consumption.
- Provide downlink carrier allocation scheme within a reasonable amount of time.

Table 2 OFDMA parameters

Number of users	10
OFDMA symbol duration	125 μs
OFDMA frame duration	5 ms
Number of data subcarriers	1024

Table 3 GA and IGA parameters

Mutation probability	0.5
Crossover operator	Single point, position changes randomly
Crossover probability	0.15
Selection operator	Roulette wheel, total replacement
Number of generations	31
Size of initial population	10

Table 4 Assumed user required rate

User id	1	2	3	4	5
Distance from base station in m	100	200	300	400	500
Requested throughput in Mbps	8	1.2	0.96	0.96	0.8
Service class (1:real-time 0:non-real-time)	0	0	0	0	0
User id	6	7	8	9	10
Distance from base station in m	100	200	300	400	500
Requested throughput in Mbps	6.4	3.2	2.32	0.88	0.88
Service class (1:real-time 0:non-real-time)	1	1	1	1	1

OFDMA parameters, IGA parameters, and users requested rate are shown in Tables 2, 3 and 4 respectively.

5.2 Channel Model

For the OFDMA channel model, we use the modified IEEE 802.16d proposed in [5] where:

$$PL_{M-IEEE} = \begin{cases} 20 \log_{10}\left(\dfrac{4\pi d}{\lambda}\right) & d \leq d_0' \\ 20 \log_{10}\left(\dfrac{4\pi d_0'}{\lambda}\right) + X & d > d_0' \end{cases} \tag{30}$$

$$X = 10\gamma \log_{10}\left(\frac{d}{d_0}\right) + C_f + C_{RX} \tag{31}$$

With d_0 is the reference distance which is 100 m. d_0' is the new reference distance which is given by:

$$d_0' = d_0 \times 10^{-\frac{C_F + C_{RX}}{10\gamma}} \tag{32}$$

C_F is the correlation coefficient for the carrier frequency f_c (MHz) and it is given by:

$$C_F = 6\log_{10}\left(\frac{f_c}{2000}\right) \tag{33}$$

C_{RX} is the correlation coefficient for the receive antenna:

$$C_{RX} = \begin{cases} -10.8\log_{10}\left(\frac{h_{RX}}{2}\right) & \text{for Type A and B} \\ -20\log_{10}\left(\frac{h_{RX}}{2}\right) & \text{for Type C} \end{cases} \tag{34}$$

where A, B and C are the terrain type which are shown in Table 5. γ is the pathloss coefficient:

$$\gamma = a - bh_{TX} + c/h_{TX} \tag{35}$$

The values of a, b and c can be taken from Table 6, given the terrain type. We have added the shadowing effect to the subcarriers in order to get a more realistic channel effect, and we have assumed that the CSI is known at the receiver in order to perform scheduling.

For our proposed fitness function A_1, A_2 and A_3 were chosen to be 1, 1 and 3 respectively.

5.3   Ordinal Optimization

As mentioned before, the objective of OO is to determine the maximum number of iterations IGA will go through before terminating its search. Since a search space of

Table 5 Types of terrain [5]

Type	Description
A	Macro-cell suburban: hilly terrain with heavy tree densities
B	Macro-cell suburban: intermediate path loss condition
C	Macro-cell suburban: flat terrain with light tree densities

Table 6 a, b and c parameters [5]

Parameter	Type A	Type B	Type C
a	4.6	4	3.6
b	0.0075	0.0065	0.005
c	12.6	17.1	20

size 10^{1024} is too large for OO, therefore a reduced search space of size 8^{12} is used. Note that it is assumed that the reduced search space has similar distribution of the entire search space; otherwise the result will be irrelevant. For the carrier allocation, OO procedure is:

1. Uniform Sampling: To uniformly sample from the reduced search space, the pseudo-code presented in Fig. 5 is used.
2. Crude Model: The crude model chosen in this case is the throughput.

$$Crude\ Model = \sum_{n=1}^{N} \sum_{k=1}^{K} C_{k,n} M_{k,n} \tag{36}$$

3. Design Parameters: The size of the good enough set chosen is G = 10 and the alignment level K = 2.
4. Order Performance Curve: The obtained curves are shown in Figs. 8 and 9. To generate the OPC, one thousand samples are chosen.
5. Selected Set Size: From the OPC obtained, the most suitable curve from Fig. 1 is bell shaped (i.e. many mediocre designs). For a bell OPC with medium noise level, we compute according to Table 1 the following:

$$Z_1 = 8.5988,\ Z_2 = 1.4089,\ Z_3 = 1.6789,\ Z_4 = 9$$

By applying Eq. 10, we obtain:

$$S = 310$$

If the initial population size is 10, and by applying Eq. 24 the maximum number of generations needed for IGA is found to be 31.

Fig. 8 Throughput versus number of occurrences

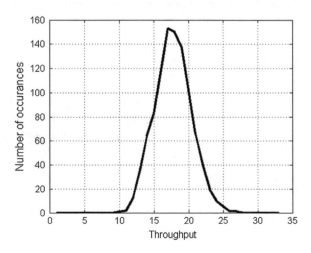

Fig. 9 Order performance curve

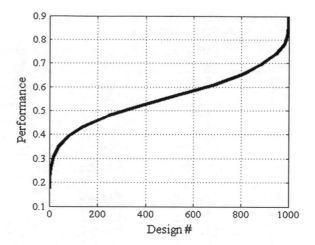

5.4 Simulation Results

Simulations were performed on an Intel(R) Core(TM) i7-2630QM CPU @ 2.00 GHz with 6 GB RAM. For the rest of the paper, all the simulation results are averaged over 1000 runs, and are executed under bad channel conditions. Other parameters are the same as in Tables 4, 5 and 6. Fitness 1, 2 and 3 are the same as in Eqs. 1, 3 and 6 respectively.

To effectively test IGA workflow, the effect of GA elements such as fitness function, mutation and initial population on the carrier allocation are first assessed followed by a complexity analysis as shown below:

1. Effect of GA operators: In this section, the effect of GA operators such as fitness function, mutation and initial population on the carrier allocation are tested. The metrics used for evaluating the performance of each algorithm are: the total throughput and its corresponding standard deviation (Table 7), the percentage of users who are under-allocated by more than 10 % (Table 8), the amount of user under-allocation (Eq. 12), and the percentage of real-time users who are under-allocated by more than 10 % (Table 8). The average results for 1000 runs are reported in Tables 7 and 8.

   a. Effect of Mutation Operator: The effect of the mutation operator is tested for IGA, with different initial population, and different fitness functions. As can be seen from Table 7, the carrier allocation with a "swap if better" mutation usually leads to a higher throughput than the carrier allocation with random mutation under the same conditions and parameter settings. On the other hand, the carrier allocation with random mutation has lower standard deviation. The amount of under-allocation and the percentage of under-allocated users are lower if "swap if better" mutation operator is used, as shown in Table 8. This is because the "swap if better" mutation either increases the

Table 7 Effect of fitness function, initial population and mutation operator on the total throughput and its corresponding standard deviation

Initial algorithm	Mutation operator	Fitness function	Total bit rate (Mbps)	Standard deviation
MRS	Swap	Fitness 1	27.60	0.24
MRS	Swap	Fitness 2	24.73	0.32
MRS	Swap	Fitness 3	24.96	0.32
MRS	Swap	Proposed	27.49	0.27
MRS	Random	Fitness 1	25.29	0.26
MRS	Random	Fitness 2	24.05	0.93
MRS	Random	Fitness 3	23.53	1.01
MRS	Random	Proposed	25.26	0.25
USIM	Swap	Fitness 1	27.75	0.41
USIM	Swap	Fitness 2	24.27	0.34
USIM	Swap	Fitness 3	24.28	0.29
USIM	Swap	Proposed	27.55	0.29
USIM	Random	Fitness 1	24.52	0.27
USIM	Random	Fitness 2	24.30	0.30
USIM	Random	Fitness 3	22.84	0.26
USIM	Random	Proposed	24.52	0.27
HA	Swap	Fitness 1	26.00	0.25
HA	Swap	Fitness 2	22.40	0.11
HA	Swap	Fitness 3	22.36	0.16
HA	Swap	Proposed	25.29	0.65
HA	Random	Fitness 1	23.20	0.22
HA	Random	Fitness 2	22.40	0.11
HA	Random	Fitness 3	22.40	0.13
HA	Random	Proposed	22.67	0.25

total throughput of the system by swapping carriers among users, or decreases the user under-allocation by donating the excess carriers to the under-allocated user; while traditional mutation operator is heuristic due to its random nature: a "good" carrier might be taken away from a user who needs it and given to a user who doesn't. Overall, the "swap if better" mutation leads to a better carrier allocation than the traditional mutation under the same test conditions.

b. Effect of Fitness Function: The effect of the fitness function is tested for IGA using "swap if better" and random mutations with different initial populations. As can be seen from Table 7, the waterfilling algorithm (fitness 1) always leads to the highest data rate, achieving 27.6, 27.75 and 26 Mbps for MRS, USIM and HA respectively. This is not surprising, as the only objective of the waterfilling algorithm is throughput maximization. Our proposed fitness function comes in second place achieving 27.49, 27.55 and

Table 8 Effect of fitness function, initial population and mutation operator on the amount of under-allocation and the percentage of under-allocated users

Initial algorithm	Mutation operator	Fitness function	Amount of under-allocation	Percentage of under-allocated users	Percentage of real-time under-allocated users
MRS	Swap	Fitness 1	0.0011	0.6	0.6
MRS	Swap	Fitness 2	0.0770	24.2	45.8
MRS	Swap	Fitness 3	0.0817	27.1	45.6
MRS	Swap	Proposed	0	0	0
MRS	Rand	Fitness 1	0.0567	25.5	40.8
MRS	Rand	Fitness 2	0.1478	27.5	43.2
MRS	Rand	Fitness 3	0.2095	27.9	38.4
MRS	Rand	Proposed	0.0542	23.5	35.2
USIM	Swap	Fitness 1	0.0186	7.8	5.6
USIM	Swap	Fitness 2	0.0701	9.7	8.2
USIM	Swap	Fitness 3	0.0678	10.1	13.2
USIM	Swap	Proposed	0	0	0
USIM	Rand	Fitness 1	0.0559	9.6	10.4
USIM	Rand	Fitness 2	0.0693	10.4	7.8
USIM	Rand	Fitness 3	0.2639	25.7	32
USIM	Rand	Proposed	0.0491	9.8	3.2
HA	Swap	Fitness 1	0.006	5	5
HA	Swap	Fitness 2	0.2951	30	40.2
HA	Swap	Fitness 3	0.2961	30.1	43.2
HA	Swap	Proposed	0.009	2.6	3
HA	Rand	Fitness 1	0.2757	28.8	36.4
HA	Rand	Fitness 2	0.2951	30	40.2
HA	Rand	Fitness 3	0.2950	30	40.2
HA	Rand	Proposed	0.2906	29.2	34

25.29 Mbps for MRS, USIM and HA respectively; which are very close to that of the waterfilling algorithm under the same conditions with comparable standard deviation, it is worthwhile noting that the slight throughput difference between our proposed fitness function and the waterfilling algorithm is compensated by a better fairness. Our proposed fitness functions has the best fairness among all algorithms. As shown in Table 8 the proposed fitness function using MRS and USIM didn't under-allocate any user through the entire simulation, while fitness 1 comes in second place, by having low amount of user-allocation of 0.0011, 0.0186 and 0.006 for MRS, USIM and HA respectively. Most importantly our proposed fitness function has the lowest percentage of under-allocated real-time users among all other fitness functions. Note that for IGA using fitness 1, the percentage of under-allocated real-time users is higher than that non-real-time users, as shown in

Table 8. This is because the waterfilling algorithm doesn't differentiate between real-time and non-real-time users. It allocates carriers based on channel condition, the non-real-time users were better off just because their requested data rate is lower than that of real-time users.

c. Effect of Initial Population: The effect of initial population is tested for IGA using "swap if better" mutation operator. In terms of total throughput as shown in Table 7, MRS and USIM performed similarly, if our proposed fitness function is used, by reaching data rate over 27 Mbps. HA performance comes in last no matter what is the fitness function. The reason for that is HA is a uniform sampling over a very large space. The initial population could not represent the distribution of the search space. Therefore HA is somewhat like a random selection that does not depend on the channel condition, while other algorithms are channel dependent that seek to maximize the throughput and fairness among users. In terms of fairness as shown in Table 8, MRS, and USIM performed the same for the proposed fitness function by meeting the QoS requirements of each and every user. For the waterfilling algorithm MRS outperformed USIM by under-allocating 0.6 % of the users and having amount of under-allocation of 0.0011. On the other hand USIM outperformed MRS for fitness 2 and fitness 3 by having lower amount of under-allocation, % of under-allocated users and % of real-time under-allocated users. That is because USIM is allocating the carriers to the users with the highest throughput, while MRS allocate the carrier to a random user who might not be the best choice. In MRS and USIM, the over-allocated user is not allocated anymore carriers once his objective is met; but since USIM has higher initial throughput, the over-allocated user is removed faster, indicating that additional carriers are allocated to the under-allocated users. Since the real-time users in this simulation have higher requested data rate, they are under-allocated in MRS more than USIM.

2. Complexity Analysis: We ran a complexity analysis using gprof (C++ profiler tool). Results are shown in Tables 9, 10 and 11.
   where:

Table 9 Profiling of the initial population selection algorithms

Function	ms/Call	# of calls	Time (ms)	Percentage time
MRS	0.07	10	0.7	4.84
USIM	0.84	10	8.4	45.9
HA	9×10^{-5}	10,000	0.9	15

Table 10 Profiling of the different fitness functions

Fitness function	Execution time (ms)
Fitness 1	1.2
Fitness 2	1.2
Fitness 3	1.8
Proposed	1.4

Table 11 Profiling of IGA for different initial population algorithms

Initial population	Execution time (ms)
IGA-MRS	6.2
IGA-USIM	18.3
IGA-HA	6

- ms/Call: represents the total time required to execute the function (including all the nested functions).
- # of calls: indicates the number of time the functions called during the execution of the program.
- % Time: shows the percentage of time it takes to run the specified function.

By examining Table 9, it can be noticed that USIM is taking long time to compute (8.4 ms). While MRS and HA are much faster by taking 0.7 and 0.9 ms to compute respectively. That is because USIM is a complex algorithms that seeks to maximize the throughput and minimize the under-allocation of the initial chromosome population.

Table 10 shows that our proposed fitness function is slightly more complex than fitness 1 and fitness 2. This is because the proposed fitness function not only computes the throughput for each user, but also the under-allocation and residue for each user. This incremental computational cost helps IGA increase the priority of under-allocated users and hence result in a "social welfare" while maintaining a fair carrier allocation for different QoS requirements.

Table 11 shows that the of IGA using MRS took 6.2 ms to execute which is very similar to that of IGA with HA, and is three times faster than IGA with USIM. The large execution time of USIM compared to MRS, and the small difference regarding the carrier allocation indicates that MRS is more suitable for real-time applications.

Note that the execution time is limited by the processor capabilities of the laptop used for this simulation, which is most likely inferior to that of a modern base station processors. Also, IGA code was executed as a single thread, had it been multi-thread we could have leveraged better multi-core and multi-thread capabilities of processors to further reduce the execution time of each proposed algorithm.

6 Conclusion

To improve on the heuristics of GA, we proposed in this paper, carrier allocation in a downlink OFDMA system by combining OO concepts and improvements to the GA standard flow. A novel fitness function is introduced: one that maximizes the throughput and minimizes the under-allocation for different types of users. OO was used to determine the maximum number of iterations for GA, while the proposed MRS, HA and USIM were suggested as alternatives to guide the selection of the initial population, and "swap if better" mutation replaced the random mutation to help IGA maximize its utility. Simulation results showed that the "swap if better" mutation operator and our proposed fitness function outperformed the traditional mutation operator and the fitness functions proposed in the literature in terms of percentage of under-allocated users, under-allocated real time users, and amount of user under-allocation, while achieving bit rate and standard deviation comparable to the waterfilling algorithm. Except for USIM, IGA was able to perform carrier allocation within few milliseconds making it suitable for real time implementation. Future work plans a parallel implementation of IGA to further improve its computational time.

References

1. A. Belghith., L. Nuaymi, Comparison of WiMAX scheduling algorithms and proposals for the rtPS QoS class, in *IEEE 14th European Wireless Conference* (2008), pp. 1–6
2. H. Cheng, W. Zhuang, Joint power-frequency-time resource allocation in clustered wireless mesh networks. Netw. IEEE **22**(1), 45–51 (2008)
3. H. Cheng, W. Zhuang, Novel packet-level resource allocation with effective qos provisioning for wireless mesh networks. IEEE Trans. Wirel. Commun. **8**(2), 694–700 (2009)
4. Y. Chiu, C. Chang, K. Feng, F. Ren, Ggra: a feasible resource-allocation scheme by optimization technique for ieee 802.16 uplink systems. IEEE Trans. Veh. Technol. **59**(3), 1393–1401 (2010)
5. Y. Cho, J. Kim, W. Yang, C. Kang, in *MIMO-OFDM Wireless Communications with MATLAB* (Wiley, New York, 2010)
6. Cisco http://www.cisco.com/en/US/solutions/collateral/ns341/ns525/ns537/ns705/ns827/white_paper_c11-520862.html (2013)
7. N. El-Zarif, M. Awad, An ordinal optimization like ga for improved ofdma system carrier allocations, in *Intelligent Systems (IS), 2012 6th IEEE International Conference* (2012), pp. 412–418
8. R. Garroppo, S. Giordano, D. Iacono, Radio-aware scheduler for wimax systems based on time-utility function and game theory, in *IEEE Global Telecommunications Conference* (2009), pp. 1–6
9. Y. Ho, R. Sreenivas, P. Vakili, Ordinal optimization of deds. Discrete Event Dyn. Syst. **2**(1), 61–88 (1992)
10. Y. Ho, Q. Zhao, Q. Jia, in *Ordinal Optimization: Soft Optimization for Hard Problems* (Springer-Verlag, New York Inc 2007)
11. R. Jayaparvathy, S. Geetha, Resource allocation and game theoretic scheduling with dynamic weight assignment in IEEE 802.16 fixed broadband wireless access systems, in *IEEE International Symposium on Performance Evaluation of Computer and Telecommunication Systems* (2008), pp. 217–224
12. S. Khemiri, G. Pujolle, K. Boussetta, N. Achir, A combined mac and physical resource allocation mechanism in ieee 802.16 e networks, in *IEEE Vehicular Technology Conference* (2010), pp. 1–5
13. T. Lau, Y. Ho, Universal alignment probabilities and subset selection for ordinal optimization. J. Optim. Theory Appl. **93**(3), 455–489 (1997)
14. Lee, L., Fang, Y, Developing a self-learning adaptive genetic algorithm, in *IEEE Proceedings of the 3rd World Congress on Intelligent Control and Automation*, vol. 1, (2002), pp. 619–624
15. Y. Luo, M. Guignard, C. Chen, A hybrid approach for integer programming combining genetic algorithms, linear programming and ordinal optimization. J. Intell. Manuf. **12**(5), 509–519 (2001)

16. M. Mehrjoo, S. Moazeni, X. Shen, A new modeling approach for utility-based resource allocation in ofdm networks, in*Proceedings of IEEE International Conference on Communications (ICC)* (2008), pp. 337–342
17. D. Niyato, E. Hossain, QoS-aware bandwidth allocation and admission control in IEEE 802.16 broadband wireless access networks: A non-cooperative game theoretic approach. Comput. Netw. **51**(11), 3305–3321 (2007)
18. D. Niyato, E. Hossain, Radio resource management games in wireless networks: an approach to bandwidth allocation and admission control for polling service in IEEE 802.16 [Radio Resource Management and Protocol Engineering for IEEE 802.16]. IEEE Wirel. Commun. **14** (1), 27–35 (2007)
19. G. Prasath, C. Fu, M. Ma, QoS scheduling for group mobility in WiMAX, in *IEEE 11th International Conference on Communication Systems* (2009), pp. 1663–1667
20. C. So-In, R. Jain, A. Tamimi, Scheduling in ieee 802.16 e mobile wimax networks: key issues and a survey. IEEE J. Sel. Areas Commun. **27**(2), 156–171 (2009)
21. J. Song, J. Li, C. Li, A cross-layer WiMAX scheduling algorithm based on genetic algorithm, in *IEEE Seventh Annual Communication Networks and Services Research Conference* (2009), pp. 292–296
22. Y. Teng, Y. Zhang, M. Song, Y. Dong, L. Wang, Genetic algorithm based adaptive resource allocation in ofdma system for heterogeneous traffic, in *Proceedings of IEEE 20th International Symposium on Personal, Indoor and Mobile Radio Communications* (2009), pp. 2060–2064
23. L. Wang, L. Zhang, D. Zheng, A class of order-based genetic algorithm for flow shop scheduling. Int. J. Adv. Manuf. Technol. **22**(11), 828–835 (2003)
24. Y. Wang, F. Chen, G. Wei, Adaptive subcarrier and bit allocation for multiuser ofdm system based on genetic algorithm, in *IEEE International Conference on Communications, Circuits and Systems* vol. 1 (2005), pp. 242–246
25. T. Wong, W. Luk, P. Heng, Sampling with Hammersley and Halton points. J. Graph. Tools **2** (2), 9–24 (1997)
26. X. Xie, H. Chen, H. Wu, Simulation studies of a fair and effective queueing algorithm for wimax resource allocation, in *IEEE Third International Conference on Communications and Networking in China* (2008), pp. 281–285
27. E. Yaacoub, Z. Dawy, Achieving the Nash bargaining solution in OFDMA uplink using distributed scheduling with limited feedback. Elsevier AEU-Int. J. Electron. Commun. (2010)
28. X. Yang, M. Venkatachalam, S. Mohanty, Exploiting the MAC layer flexibility of WiMAX to systematically enhance TCP performance, in *IEEE Mobile WiMAX Symposium* (2007), pp. 60–65
29. Y. Yu, W. Zhou, Resource allocation for ofdma system based on genetic algorithm, in *IEEE International Workshop on Cross Layer Design* (2007), pp. 65–69

Structure-Oriented Techniques for XML Document Partitioning

Gianni Costa and Riccardo Ortale

Abstract Focusing on only one type of structural component in the process of clustering XML documents may produce clusters with a certain extent of inner structural inhomogeneity, due either to uncaught differences in the overall logical structures of the available XML documents or to inappropriate choices of the targeted structural component. To overcome these limitations, two approaches to clustering XML documents by multiple heterogeneous structures are proposed. An approach looks at the simultaneous occurrences of such structures across the individual XML documents. The other approach instead combines multiple clusterings of the XML documents, separately performed with respect to the individual types of structures in isolation. A comparative evaluation over both real and synthetic XML data proved that the effectiveness of the devised approaches is at least on a par and even superior with respect to the effectiveness of state-of-the-art competitors. Additionally, the empirical evidence also reveals that the proposed approaches outperform such competitors in terms of time efficiency.

Keywords Data mining · XML clustering · XML transactional representation · Ensemble XML clustering

1 Introduction

The eXtensible Markup Language[1] (XML) [1, 2] is a standard for representing, storing and exchanging data in a human-intelligible and machine-readable textual format, wherein pure content is given a convenient logical structure. The latter helps

[1]http://www.w3c.org.

G. Costa · R. Ortale (✉)
ICAR-CNR, Via P. Bucci 41C, 87036 Rende, CS, Italy
e-mail: ortale@icar.cnr.it

G. Costa
e-mail: costa@icar.cnr.it

© Springer International Publishing Switzerland 2016   167
M. Hadjiski et al. (eds.), *Novel Applications of Intelligent Systems*,
Studies in Computational Intelligence 586, DOI 10.1007/978-3-319-14194-7_9

explaining the nested content, thus being useful for better information discovery and management.

XML data is a challenging research domain [3], that calls for suitable methods for information handling [4, 5]. Indeed, the structural information of XML data is not considered by traditional approaches [6, 7] to information handling, which are either devoted to the management of highly structured data, such as relational databases, or too focused on the textual nature of the underlying data, such as in the case of information retrieval techniques.

Catching structural resemblance between XML documents enables more effective processing of XML data [5, 8–11]. In particular, clustering XML documents by structure may be useful in several applicative scenarios. It can support the extraction of (schema or DTD) structures from a collection of XML documents [12, 13] or help to identify different sources providing the same kind of information [14]. Moreover, detecting structurally-homogeneous clusters of XML documents helps in devising indexing techniques for such documents, thus improving XML query formulation and optimization [15].

However, two main issues arise in clustering XML documents by their structural components. Firstly, XML data can share various forms of common structural features (e.g., nodes, edges, paths [16–19], (sub)trees [15, 20–22], s-graphs [23], summaries [23] and so forth) and, generally, the most appropriate one to be chosen for effective clustering is not known beforehand. Secondly, judging differences only in terms of one type of structural components may not suffice to effectively separate the available XML documents. In fact, a careful investigation of the resulting clusters is likely to reveal an extent of intra-cluster inhomogeneity, that may be due to some uncaught differences in the structures of the XML documents within the same clusters, ascribable to further unconsidered forms of structural components.

The above issues suggest that the task of clustering XML documents by structure should be settled in a more general framework, that allows both to accommodate disparate types of structural components and to assess the impact of their discriminative power on clustering effectiveness.

The potential of using multiple types of structural components for XML document clustering was preliminarily studied in [24], with a different purpose, i.e., the generation of a hierarchy of nested clusters. The aim of this chapter is instead to explore the exploitation of multiple types of substructures for developing partitioning approaches to XML document clustering. To this end, a transactional framework that can accommodate all types of tree-like substructures within the available XML documents is used to allow the user to specify the most appropriate (combination of) structural components to be looked at in the clustering process.

Based on the underlying transactional framework, two XML clustering approaches are devised. XPMCS (*XML Partitioning based on Multiple Co-occurrent Substructures*) [25] represents each XML document as a transaction over the chosen combination of substructures and looks at their simultaneous co-occurrence across the resulting transactions, in order to isolate structurally-homogeneous clusters of XML documents. Instead, the other approach, referred to as XPCMC (*XML Partitioning based on Combining Multiple Clusterings*), separately represents the available XML documents as transactions over each chosen type of substructures in isolation. The distinct transactional representations of the original XML documents are then independently clustered and the results of the multiple clusterings are eventually combined into one final partition.

Both the devised approaches are autonomous techniques, in the sense that no further user intervention is required, apart from the initial specification of the types of the structural components to be addressed for clustering. The specification (of a combination) of structural components allows the embedding of human knowledge on both the specific XML domain and the semantics of the clustering application. Such a knowledge is necessary to capture different possible explanations of the XML data at hand (according to the pursued clustering task) and can be provided by domain analysts, fixed classification criteria, or in-depth customized analysis.

A comparative evaluation over both real and artificial XML data reveals that the proposed approaches are more efficient and at least as effective as established competitors.

The outline of this chapter is as follows. Section 2 introduces notation and preliminaries. Sections 3 and 4 treat, respectively, the XPMCS and XPCMC approaches. Section 5 presents experimental evaluation of the devised techniques. Finally, Sect. 6 concludes and highlights major directions of future research.

2 Preliminaries

Before delving into the details of our techniques, we preliminarily introduce the notation used throughout the chapter as well as some basic concepts. The structure of XML documents without references can be modeled in terms of *rooted labeled trees*, that represent the hierarchical relationships among the document elements (i.e., nodes).

Definition 1 (*XML Tree*) An XML tree is a rooted, labeled tree, represented as a tuple $\mathbf{t} = (r_{\mathbf{t}}, \mathbf{V_t}, \mathbf{E_t}, \lambda_{\mathbf{t}})$, whose individual components have the following meaning:

- $\mathbf{V_t}$ is a set of nodes;
- $\mathbf{E_t} \subseteq \mathbf{V_t} \times \mathbf{V_t}$ is a set of edges, catching the parent-child relationships between nodes of $\mathbf{t}$;
- $r_{\mathbf{t}} \in \mathbf{V_t}$ is the root node of $\mathbf{t}$, i.e. the only node with no entering edges;
- Σ is an alphabet of node tags (i.e., labels);
- $\lambda_t : \mathbf{V_t} \mapsto \Sigma$ is a node labeling function.

In the above definition, both the elements of XML documents and their attributes are mapped to nodes in the corresponding XML-tree representation.

Let n_i and n_j be two nodes from $\mathbf{V_t}$. n_i is the parent of n_j (and, dually, n_j is a child of n_i), if $(n_i, n_j) \in \mathbf{E_t}$. This type of parent-child hierarchical relationship is represented as $n_i \rightarrow n_j$. Instead, if there is a path from n_i to n_j of any positive length

p (representing intermediate edges), n_i is an ancestor of n_j, whereas n_j is a descendant of n_i. The ancestor-descendant hierarchical relationship is indicated as $n_i \xrightarrow{p} n_j$: clearly, if $p = 1$, the ancestor-descendant relationship reduces to the parent-child relationship $n_i \to n_j$. The set of all paths from r_t to any node n in $\mathbf{V_t}$ is denoted as $paths(\mathbf{t})$, i.e., $paths(\mathbf{t}) = \{r_t \xrightarrow{p} n | r_t, n \in \mathbf{V_t}, p \geq 1\}$.

Nodes in $\mathbf{V_t}$ divide into two disjoint subsets: the set $\mathbf{L_t}$ of *leaves* and the set $\mathbf{V_t} - \mathbf{L_t}$ of *inner nodes*. An inner node has at least one child. There is a (predefined) left-to-right ordering among the siblings of each inner node [1]. A leaf is instead a node with no children.

All nodes from $\mathbf{V_t}$ can be numbered according to their position in the pre-order depth-first traversal of $\mathbf{t}$. Assume that function $num_t : \mathbf{V_t} \mapsto \mathbf{N}$ maps nodes to their numbering and that m_i and m_j are two sibling nodes from $\mathbf{V_t}$. Notation $num_t(m_i) < num_t(m_j)$ indicates that m_i precedes m_j in the (predefined) sibling ordering.

Tree-like structures are also used to represent generic structural patterns occurring across a collection of XML trees (such as individual nodes, edges as well as paths).

Definition 2 (*Substructure*) Let $\mathbf{t}$ and $\mathbf{s}$ be two XML trees. $\mathbf{s}$ is a substructure of $\mathbf{t}$, if there exists a total function $\varphi : \mathbf{V_s} \to \mathbf{V_t}$, that satisfies the following conditions for each $n, n_i, n_j \in \mathbf{V_s}$:

- $(n_i, n_j) \in \mathbf{E_s}$ iff $\varphi(n_i) \xrightarrow{p} \varphi(n_j)$ in $\mathbf{t}$ with $p \geq 1$;
- $num_s(n_i) < num_s(n_j)$ iff $num_t(\varphi(n_i)) < num_t(\varphi(n_j))$;
- $\lambda_s(n) = \lambda_t[\varphi(n)]$.

The mapping φ preserves node labels and hierarchical relationships. In this latter regard, depending on the value of p, two definitions of substructures can be distinguished. In the simplest case $p = 1$ and a substructure $\mathbf{s}$ is simply an *induced* tree pattern that matches a contiguous portion of t, since φ maps the parent-child edges of $\mathbf{s}$ onto parent-child edges of t. This is indicated as $\mathbf{s} \sqsubseteq \mathbf{t}$. A more general definition follows when $p \geq 1$ [26]. In such a case, $\mathbf{s}$ matches not necessarily contiguous portions of $\mathbf{t}$, since φ summarizes hierarchical relationships by mapping parent-child edges of $\mathbf{s}$ into either parent-child or ancestor-descendant edges of $\mathbf{t}$. This is denoted as $\mathbf{s} \preceq \mathbf{t}$ and $\mathbf{s}$ is also said to be an *embedded* tree pattern of $\mathbf{t}$.

Hereafter, the notions of substructure, (structural) component and tree pattern are used as synonyms.

Clustering by structure aims to divide a collection $\mathcal{D} = \{\mathbf{t_1}, \ldots, \mathbf{t_N}\}$ of N XML trees to form a partition $\mathcal{P} = \{\mathcal{C}_1, \ldots, \mathcal{C}_K\}$ of nonempty clusters such that $\mathcal{C}_i \subseteq \mathcal{D}$ and $\mathcal{C}_i \cap \mathcal{C}_j = \emptyset$ for all $i, j = 1, \ldots, K$ with $i \neq j$. The clustering process generally attempts to maximize the degree of structural homogeneity exhibited by the XML trees in the same cluster and to minimize the extent of structural homogeneity between XML trees within distinct clusters.

In this chapter, we develop two parameter-free clustering approaches accounting for various forms of structural components, that differentiate the available XML trees.

All types of tree-like components can be handled for clustering. Notation $\mathcal{T}^{(\cdot)}$ is used to denote one generic type of structural component addressed for clustering. Example definitions of $\mathcal{T}^{(\cdot)}$ include:

- The selection of one-node substructures modeling the individual nodes in the XML trees, i.e.,

$$\mathcal{T}^{(n)} = \{s : \mathbf{V}_s = \{r_s\}, \exists t \in \mathcal{D}, s \sqsubseteq t\}$$

  In such a case, the generic substructure s consists only of its root r_s, that matches some corresponding node of an XML tree t in $\mathcal{D}$.
- The selection of one-edge substructures modeling the parent-child relationships in the XML trees, i.e.,

$$\mathcal{T}^{(e)} = \{s : \mathbf{V}_s = \{r_s, n\}, \mathbf{E}_s = \{(r_s, n)\}, \exists t \in \mathcal{D}, s \sqsubseteq t\}$$

  Here, the individual substructure s consists only of one edge (r_s, n), that matches some corresponding parent-child edge of an XML tree t in $\mathcal{D}$.
- The selection of one-path substructures modeling the distinct root-to-node paths in the XML trees, i.e.,

$$\mathcal{T}^{(p)} = \{s : |paths(s)| = 1, \exists t \in \mathcal{D}, s \sqsubseteq t\}$$

  Each s in the above set $\mathcal{T}^{(p)}$ is hence linear, i.e., consists of only one path matching some (root-to-node or root-to-leaf) path of an XML tree t in $\mathcal{D}$.

Interestingly, more sophisticated definitions of $\mathcal{T}^{(\cdot)}$ may involve induced or embedded subtrees.

In the rest of this section, the generic set of substructures used for clustering is abstractedly denoted as $\mathcal{R}$ for the purpose of formalizing how the transactional framework is used in the context of the XPMCS and XPCMC approaches. Details on the actual definition of $\mathcal{R}$ shall be provided in Sect. 5.

The substructures in $\mathcal{R}$ enable the projection of the original XML trees into a high-dimensional space, wherein the occurrence of the individual substructures within each XML tree is explicitly represented. More precisely, the XML trees can be modeled as transactions over a feature space $\mathcal{S} = \{\mathcal{F}_s | s \in \mathcal{R}\}$. Here, the generic feature $\mathcal{F}_s$ is a Boolean attribute, that indicates the presence/absence of the related component s of $\mathcal{R}$ in the individual XML trees.

Let $\mathbf{x}^{(t)}$ be the high-dimensional transactional representation over $\mathcal{S}$ of an XML tree t. The value of each attribute $\mathcal{F}_s$ in the context of $\mathbf{x}^{(t)}$ is true if s is a substructure of t, otherwise it is false. Hence, $\mathbf{x}^{(t)}$ can be modeled as a proper subset of

$\mathcal{S}$, namely $\mathbf{x}^{(t)} = \{\mathcal{F}_\mathbf{s} \in \mathcal{S}|\mathbf{s}\sqsubseteq\mathbf{t}\}$, with the meaning that the features explicitly present in $\mathbf{x}^{(t)}$ take value true, whereas the value of the missing ones is false.

The transactional representation of the XML trees is beneficial for both the devised approaches. Indeed, in both XPMCS and XPCMC, the cost for testing the presence of the selected components within the transactions associated with the XML trees is independent of the structural complexity of the same components. Nonetheless the transactional representation involves the non-trivial discovery of meaningful clusters in large-scale databases of high-dimensional transactions. From this perspective, the XPMCS and XPCMC approaches reformulate the original problem of grouping $\mathcal{D}$ by structure as that of finding structurally-pure clusters in the transactional representations of $\mathcal{D}$ over the (selected types of) structural components.

3 The XPMCS Approach

The basic idea behind cluster separation in the transactional framework consists in projecting an input set $\mathcal{D}$ of XML trees into a high-dimensional feature space $\mathcal{S}$, in which to isolate homogeneous groups of transactions sharing discriminatory co-occurrences of structural features. In the context of the XPMCS approach, the feature space $\mathcal{S}$ is built over an underlying collection $\mathcal{R}$, that is in turn a collection of heterogeneous types of substructures.

The scheme of the proposed approach is depicted in Fig. 1. It initially projects the individual XML trees within the input forest $\mathcal{D}$ into the space $\mathcal{S}$ of clustering features. Such a projection results in the transactional representation $\mathbf{D}$ of the original forest $\mathcal{D}$.

Finding clusters in the high-dimensional feature space $\mathcal{S}$ is problematic for various reasons [27]. Primarily, transactions tend to form different clusters on distinct subsets of features, which penalizes the effectiveness of clustering and exacerbates its time requirements. Secondarily, poor scalability with both the size and the dimensionality of transactions is usually a major limitation. Yet, an underestimation (resp. overestimation) of the number of child clusters to isolate in the input set $\mathcal{D}$ misses (resp. uncovers) actual (resp. artificial) groups.

```
XPMCS(𝒟)
   Input:  a forest 𝒟 = {t₁,...,t_N} of XML trees;
           the set of selected substructures ℛ;
   Output: a partition 𝒫 of 𝒟;
   1:  let 𝒮 ← {ℱ_s|s ∈ ℛ} be the feature space
   2:  let x^(t_i) ← {ℱ_s ∈ 𝒮|s ⊑ t_i} for each i = 1,...,N;
   3:  let D ← {x^(t_i) ⊆ 𝒮|t_i ∈ 𝒟};
   4:  𝒫 ← Generate-Clusters(D);
   5:  RETURN 𝒫;
```

Fig. 1 The scheme of the XPMCS clustering approach

To deal with specificities of the transactional setting, the XML trees within the input set $\mathcal{D}$ are separated through the GENERATE-CLUSTERS algorithm proposed in [27]. The latter is an effective and parameter-free technique for transactional clustering, that automatically partitions $\mathcal{D}$ into an appropriate number of child clusters.

The fundamentals of GENERATE-CLUSTERS are reviewed below for the sake of self-containment. Instead, a discussion on the convergence of GENERATE-CLUSTERS along with a comparative analysis of its empirical behavior against a wide variety of established competitors can be found in [27].

The general scheme of the GENERATE-CLUSTERS algorithm is reported in Fig. 2. The algorithm starts with a partition $\mathcal{P}$ containing a single cluster corresponding to the whole transactional dataset **D** (line L1). (line L1). The core of the algorithm is the body of the loop between lines L2–L15. Within the loop, an attempt to generate a new cluster is performed by (*i*) choosing a candidate node (corresponding to a cluster with low quality) to split (line L4); (*ii*) splitting the candidate cluster into two child clusters (line L5); and (*iii*) evaluating whether the splitting allows a new partition exhibit better quality than the original partition (lines L6–L13). If this is the case, the loop can be stopped (line L10) and the partition is updated, by replacing the candidate cluster with the new child clusters (line L8). Viceversa, child clusters are discarded and a new candidate cluster is considered for splitting.

The PARTITION-CLUSTER procedure at line L5 iteratively evaluates, for each transaction $\mathbf{x}^{(\mathbf{t})} \in \mathcal{C}_i \cup \mathcal{C}$, whether a membership reassignment improves the degree of structural homogeneity of the two clusters. The contribution of $\mathbf{x}^{(\mathbf{t})}$ to structural homogeneity is evaluated in two cases: both in the case that $\mathbf{x}^{(\mathbf{t})}$ is maintained in its original cluster of membership and in the case that $\mathbf{x}^{(\mathbf{t})}$ is moved to the other cluster. If moving $\mathbf{x}^{(\mathbf{t})}$ causes an improvement in the structural homogeneity, then the swap is accepted.

Fig. 2 The GENERATE-CLUSTERS procedure

```
GENERATE-CLUSTERS(D)
    Input: A set D = {x^(t1),...,x^(tN)} of transactions corre-
        sponding to
            XML trees;
    Output: A partition P = {C1,...,Ck} of clusters of transac-
        tions
            corresponding to XML trees;
    L1:  let P ← {D};
    L2:  repeat
    L3:     Generate a new cluster C of transactions, initially
            empty;
    L4:     for each cluster Ci ∈ P do
    L5:        PARTITION-CLUSTER(Ci, C);
    L6:        P' ← P ∪ {C};
    L7:        if Quality(P) < Quality(P') then
    L8:           P ← P';
    L9:           STABILIZE-CLUSTERS(P);
    L10:             break
    L11:        else
    L12:           Restore all x^(tj) ∈ C into Ci;
    L13:        end if
    L14:     end for
    L15:  until no further cluster C can be generated
    L16:  RETURN P;
```

The local quality $Quality(\mathcal{C})$ of a cluster $\mathcal{C}$ is a key component of GENERATE-CLUSTERS and measures the degree of structural homogeneity within $\mathcal{C}$. More precisely, $Quality(\mathcal{C})$ is defined as the gain in feature strength with respect to the whole transactional dataset $\mathbf{D}$, i.e.,

$$Quality(\mathcal{C}) = \Pr(\mathcal{C}) \sum_{\mathcal{F} \in \mathcal{S}_\mathcal{C}} \left[\Pr(\mathcal{F}|\mathcal{C})^2 - \Pr(\mathcal{F}|\mathbf{D})^2 \right]$$

where $\Pr(\mathcal{F}|\mathcal{C})^2$ corresponds to the relative strength of $\mathcal{F}$ within $\mathcal{C}$, whereas $\Pr(\mathcal{C})$ represents the relative strength of $\mathcal{C}$. These two factors work in contraposition: singleton clusters exhibit strong features in a sparse region, whereas highly populated clusters exhibit weaker features in a dense region. The above formula finds an interpretation in terms of subspace clustering. Features exhibiting a high occurrence frequency with respect to the occurrence frequency in the whole dataset $\mathbf{D}$, define a subset of relevant features, as opposed to low-occurrence features which are indeed irrelevant for the purpose of clustering. Thus, clusters exhibit high quality whenever a subset of relevant features occurs, whose frequency is significantly higher than in the whole dataset $\mathbf{D}$.

Differently from the PARTITION-CLUSTER procedure, where the improvement in quality is attempted locally to a cluster, the STABILIZE-CLUSTERS procedure tries to increase the global partition quality $Quality(\mathcal{P})$. This is accomplished by finding, for each transaction, the most suitable cluster among the ones available in the partition. The quality $Quality(\mathcal{P})$ of a partition $\mathcal{P}$ is meant to measure both the homogeneity of clusters and their compactness. Viewed in this respect, partition quality is defined as follows

$$Quality(\mathcal{P}) = \sum_{\mathcal{C} \in \mathcal{P}} \Pr(\mathcal{C})\, Quality(\mathcal{C})$$

Notice that the component $Quality(\mathcal{C})$ is already proportional to the contribution $\Pr(\mathcal{C})$. As a result, in the overall partition quality, the contribution of each cluster is weighted by $\Pr(\mathcal{C})^2$. This weighting has a major effect in the GENERATE-CLUSTERS procedure: splitting in very small clusters is penalized. Indeed, the generated clusters are added to the partition only if their contribution is really worth.

4   The XPCMC Approach

The previous approach finds clusters of XML trees by looking at the co-occurrence of multiple types of substructures within their logical structures.

An alternative consists in separately performing multiple clusterings of the same XML documents with respect to each type of substructures in isolation and then refining the resulting partitions through a combined clustering. This process is

Fig. 3 The scheme of the XPCMC clustering approach

XPCMC($\mathcal{D}$)
Input: a forest $\mathcal{D} = \{\mathbf{t}_1, \ldots, \mathbf{t}_N\}$ of XML trees;
        the set of selected substructures $\mathcal{R}$;
Output: a partition $\mathcal{P}$ of $\mathcal{D}$;
P1: let k be the number of types of substructures chosen for clustering;
P2: let $\mathcal{S}_l \leftarrow \{\mathcal{F}_\mathbf{s} | \mathbf{s} \in \mathcal{R}_l\}$ be the l-th feature space, for each $l = 1, \ldots, k$;
P3: $\mathbf{x}_l^{(\mathbf{t}_i)} \leftarrow \{\mathcal{F}_\mathbf{s} \in \mathcal{S}_l | \mathbf{s} \sqsubseteq \mathbf{t}_i\}$ for each $l = 1, \ldots, k$ and for each $i = 1, \ldots, N$;
P4: $\mathbf{D}^{(l)} \leftarrow \{\mathbf{x}_l^{(\mathbf{t}_i)} \subseteq \mathcal{S}_l | \mathbf{t}_i \in \mathcal{D}\}$ for each $l = 1, \ldots, k$;
P5: $\mathcal{P}^{(l)} \leftarrow$ GENERATE-CLUSTERS($\mathbf{D}^{(l)}$) for each $l = 1, \ldots, k$;
P6: /* build one partition $\mathcal{P}$ of $\mathcal{D}$ by combining $\mathcal{P}^{(1)}, \ldots, \mathcal{P}^{(k)}$ according to the scheme in [28], which is algorithmically described by the below lines;*/
P7: $\mathcal{S} = \{\mathcal{F}_{\mathcal{C}_h^{(l)}} | l = 1, \ldots, k \wedge h = 1, \ldots, |\mathcal{P}^{(l)}| \wedge \mathcal{C}_h^{(l)} \in \mathcal{P}^{(l)}\}$;
P8: $\mathbf{x}^{(\mathbf{t}_i)} \leftarrow \{\mathcal{F}_{\mathcal{C}_h^{(l)}} \in \mathcal{S} | \mathbf{t}_i \in \mathcal{C}_h^{(l)}\}$ for each $i = 1, \ldots, N$;
P9: $\mathbf{D} \leftarrow \{\mathbf{x}^{(\mathbf{t}_i)} \subseteq \mathcal{S} | \mathbf{t}_i \in \mathcal{D}\}$;
P10: $\mathcal{P} \leftarrow$ GENERATE-CLUSTERS($\mathbf{D}$);
P11: RETURN $\mathcal{P}$;

implemented by the XPCMC approach, that can be viewed as an ensemble-clustering instance, aimed to project the original XML trees from $\mathcal{D}$ into distinct spaces of boolean features. Formally, the chosen collections of substructures $\mathcal{R}_1, \ldots, \mathcal{R}_k$ (where k is the overall number of substructure types) originate as many feature spaces $\mathcal{S}_1, \ldots, \mathcal{S}_k$, into which the XML trees at hand are projected to produce the transactional representations $\mathbf{D}^{(1)}, \ldots, \mathbf{D}^{(k)}$. These representations are then independently clustered over their respective feature spaces $\mathcal{S}_1, \ldots, \mathcal{S}_k$ and the resulting partitions are eventually combined into one partition.

The scheme of the XPCMC approach is sketched in Fig. 3. The algorithm begins by building the feature spaces $\mathcal{S}_l$ from the corresponding sets $\mathcal{R}_l$ of structural components (line P2). Through projections in the feature spaces $\mathcal{S}_l$ (line P3), the distinct transactional representations $\mathbf{D}^{(l)}$ of $\mathcal{D}$ are produced (line P4). Such transactional representations are then separately clustered (at line P5) by means of the GENERATE-CLUSTERS procedure (reviewed in Sect. 3). This results into as many partitions $\mathcal{P}^{(1)}, \ldots, \mathcal{P}^{(k)}$ of $\mathcal{D}$, that are finally combined into one partition $\mathcal{P}$ through the scheme proposed in [28]. Such a scheme consists in another round of transactional clustering performed over the space of boolean features induced by the multiple partitions $\mathcal{P}^{(1)}, \ldots, \mathcal{P}^{(k)}$. Let $|\mathcal{P}^{(l)}|$ denote the number of clusters within partition $\mathcal{P}^{(l)}$. Also, assume that $\mathcal{C}_h^{(l)}$ corresponds to the hth cluster of partition $\mathcal{P}^{(l)}$, with h ranging from 1 to $|\mathcal{P}^{(l)}|$. The scheme in [28] for the combination of multiple clusterings can be formalized in the context of the XPCMC approach and its underlying transactional framework as a further transactional clustering, in which the XML trees in $\mathcal{D}$ are projected (at line P8) into the feature space $\mathcal{S} = \{\mathcal{F}_{\mathcal{C}_h^{(l)}} | l = 1, \ldots, k \wedge h = 1, \ldots, |\mathcal{P}^{(l)}| \wedge \mathcal{C}_h^{(l)} \in \mathcal{P}^{(l)}\}$ (built at line P7). By paralleling the discussion in Sect. 2, such a projection enables the representation of each XML tree $\mathbf{t}_i \in \mathcal{D}$ as a transaction $\mathbf{x}^{(\mathbf{t}_i)} = \{\mathcal{F}_{\mathcal{C}_h^{(l)}} \in \mathcal{S} | \mathbf{t}_i \in \mathcal{C}_h^{(l)}\}$ grouping the k clusters, into which the XML tree $\mathbf{t}_i$ was placed by the separate clustering processes (at line P5).

To this point, the transactional representation **D** (resulting at line P9) is clustered in order to combine the previous partitions (discovered at line P5) into one partition $\mathcal{P}$. Notice that the exploitation (at line P10) of the GENERATE-CLUSTERS procedure for this latter purpose deviates from the scheme in [28] and is deliberately chosen to avoid a critical problem, i.e., the specification of the number of clusters for the final partition $\mathcal{P}$.

The equivalence relation between consensus partition and transactional clustering over a space of binary features was established in [28].

To the best of our knowledge, XPCMC is the first approach to structure-oriented XML partitioning based on combining multiple clusterings independently performed to separately operate on various substructures of the available XML trees.

5 Evaluation

In this section, the behavior of the devised clustering approaches is comparatively evaluated against various competitors across different datasets. All experiments were conducted on a Linux machine, equipped with an Intel Core i5 processor, 2.30 GHz of clock speed and 4 GB of memory.

5.1 Data Sets

Standard benchmark data sets were employed for a direct comparison against the competitors. These data sets are described below.

We choose two real-world data sets, characterized by imbalanced distributions of the classes of XML documents.

DBLP is a bibliographic archive of scientific publications on computer science (http://dblp.unitrier.de/xml/). The archive is available as one very large XML file with a diversified structure. The whole file is decomposed into 479,426 XML documents corresponding to as many scientific publications. These individually belong to one of 8 classes: article (173,630 documents), proceedings (4764 documents), mastersThesis (5 documents), incollection (1379 documents), inproceedings (298,413 documents), book (1125 documents), www (38 documents), phdthesis (72 documents). The individual classes of XML documents exhibit differentiated structures, despite some overlap among certain document tags (such as *title*, *author*, *year* and *pages*), that occur in (nearly) all of the XML documents.

The *Sigmod* collection groups 988 documents complying to three different class DTDs: IndexTermsPage, OrdinaryIssue and Proceedings. These classes contain, respectively, 920, 51 and 17 XML documents. Such classes have diversified structures, despite the occurrence of some overlapping tags, such as *volume*, *number*, *authors*, *title* and *year*.

Three synthetic data sets were generated from as many collections of DTDs.

The first synthesized data set, referred to as *Synth1*, comprises 1000 XML documents produced from a collection of 10 heterogeneous DTDs (illustrated in Fig. 6 of [3]), that were individually used to generate 100 XML documents. These DTDs exhibit strong structural differences and, thus, can be neatly separated by most clustering algorithms.

A finer evaluation can be obtained by investigating the behavior of the compared algorithms on a collection of XML documents, that are very similar to one another from a structural point of view. To perform such a test, a second synthesized data set, referred to as *Synth2* and consisting of 3000 XML documents, was assembled from 3 homogeneous DTDs (illustrated in Fig. 7 of [3]), individually used to generate 1000 XML documents. Experiments over *Synth2* clearly highlight the ability of the competitors at operating in extremely-challenging applicative-settings, wherein the XML documents share multiple forms of structural patterns.

Finally, *Synth3* consists of 1400 synthesized documents individually belonging to one of 7 distinct class DTDs. These classes represent a challenging domain for the approaches to structural classification, since they were suitably designed in [15] to overlap in at most 30 % of their element definitions. This implies a commonality of nodes, edges and even paths in the documents conforming to the different classes.

5.2 Competitors

The behavior of XPMCS and XPCMC was compared over the chosen data sets against a selection of state-of-the-art competitors, namely *SGrace* [23], *XRep* [15] and *XProj* [3].

S-GRACE is a hierarchical clustering algorithm, that groups XML documents according to the structural information in their associated *s-graphs*. The notion of s-graph denotes a minimal summary of edge containment in one or multiple XML trees. More precisely, given a cluster of XML trees, the associated s-graph is a directed graph whose nodes and edges correspond, respectively, to the nodes (as well as attributes) and parent-child edges of the XML trees in the cluster. The distance between two clusters is defined as the proportion of non-overlapping edges in the associated s-graphs. *XRep* is an adaptation of the agglomerative hierarchical algorithm. Initially, each XML tree is placed in its own cluster. The algorithm then walks into an iterative step, in which the least dissimilar clusters are merged. Cluster merging halts when an optimal partition (i.e. a partition whose intra-distance within clusters is minimized and inter-distance between clusters is maximized) is reached. *XProj* is a partitioning method that initially divides the available XML documents into k random clusters. Each such a cluster is equipped with a representative, i.e., a collection of substructures with a fixed number n of nodes, that frequently occur in the cluster. Henceforth, the algorithm reiterates the relocation of the XML documents. These are individually reassigned to the clusters with the best matching representative substructures. Each single relocation is followed by the re-computation of the frequent representative substructures within the individual clusters. Relocation eventually halts when either the average intra-cluster structural cohesiveness does not significantly increase or a (pre-specified) maximum number of relocations has been performed.

In particular, in our experimental analysis, the results of the comparisons against *SGrace* [23] and *XRep* [15] are obtained by exploiting our implementations of such competitors. Instead, the comparison against *XProj* is indirect, i.e. based on the results reported in [3]. Notwithstanding, the adoption of the same data sets used in [3], namely *Sigmod*, *Synth*1 and *Synth*2, still enables a meaningful comparison against *XProj*.

5.3   Empirical Results

We resort to external criteria to evaluate clustering effectiveness. More specifically, we investigate the behavior of XPMCS and XPCMC over collections of XML documents with known class labels and analyze the correspondence between the discovered and hypothesized structures. Clustering effectiveness is measured in terms of precision and recall [29]. A partition $\mathcal{P} = \{\mathcal{C}_1, \ldots, \mathcal{C}_k\}$ of $\mathcal{D}$ can be summarized into a contingency table m, where columns represent discovered clusters and rows represent true classes. Each entry m_{ij} indicates the number of transactions (corresponding to XML documents in $\mathcal{D}$), that are assigned to cluster $\mathcal{C}_j$, with $1 \leq j \leq k$, and actually belong to class $\mathbf{C}_i$, with $1 \leq i \leq t$. Intuitively, each cluster $\mathcal{C}_j$ corresponds to the class $\mathbf{C}_i$ that is best represented in $\mathcal{C}_j$, i.e., such that m_{ij} is maximal. For any cluster $\mathcal{C}_j$, the index $h(j)$ of the class $\mathbf{C}_i$ with maximal m_{ij} is defined as $h(j) = max_i\ m_{ij}$. Precision $P(\mathcal{C}_j)$ and recall $R(\mathcal{C}_j)$ for the generic cluster $\mathcal{C}_j$ are defined as follows:

$$P(\mathcal{C}_j) = \frac{|\{\mathbf{x}^{(\mathbf{t})} \in \mathcal{C}_j | \mathbf{t} \in \mathbf{C}_{h(j)}\}|}{|\mathcal{C}_j|}, \quad R(\mathcal{C}_j) = \frac{|\{\mathbf{x}^{(\mathbf{t})} \in \mathcal{C}_j | \mathbf{t} \in \mathbf{C}_{h(j)}\}|}{|\mathbf{C}_{h(j)}|}$$

The average precision P and recall R for the whole partition $\mathcal{P}$ can be defined as

$$P = \frac{1}{|\mathcal{P}|}\sum_{\mathcal{C} \in \mathcal{P}} P(\mathcal{C}), \quad R = \frac{1}{|\mathcal{P}|}\sum_{\mathcal{C} \in \mathcal{P}} R(\mathcal{C})$$

Table 1 shows the results of the comparative evaluation of the proposed approaches, when both operate on the XML document nodes, edges and root-to-leaf paths. Precisely, XPMCS operates on the chosen substructures simultaneously, i.e., $\mathcal{R} = \mathcal{T}^{(n)} \cup \mathcal{T}^{(e)} \cup \mathcal{T}^{(p)}$. Instead, in XPCMC, the separate sets $\mathcal{T}^{(n)}, \mathcal{T}^{(e)}$ and $\mathcal{T}^{(p)}$ of substructures are respectively addressed in the context of three independent

Table 1 Summary of comparative evaluation

Collection	No. of docs	Classes	Method	Clusters	Precision	Recall	Time (s)
DBLP	479,426	8	XPMCS	8	0.97	0.92	1983
			XPCMC	8	0.98	0.97	2742
			S-Grace	–	–	–	–
			XRep	–	–	–	–
Sigmod	998	3	XPMCS	3	1	1	6
			XPCMC	3	3	3	14
			S-Grace	3	0.67	0.88	58
			XRep	3	1	1	186
Synth1	1000	10	XPMCS	10	1	1	12
			XPCMC	10	1	1	19
			S-Grace	10	1	1	49
			XRep	11	0.91	0.96	318
			XProj	10	1	1	–
Synth2	3000	3	XPMCS	3	1	1	7
			XPCMC	3	1	1	13
			S-Grace	3	0.99	0.99	1151
			XRep	3	0.67	0.93	2172
	300	3	XProj	3	1	1	–
Synth3	1400	7	XPMCS	7	1	1	5
			XPCMC	7	1	1	8
			S-Grace	7	1	1	92
			XRep	7	0.85	0.97	146

clusterings of each XML corpus, whose results are eventually combined (recall that the entities $\mathcal{R}, \mathcal{T}^{(n)}, \mathcal{T}^{(e)}$ and $\mathcal{T}^{(p)}$ were introduced in Sect. 2).

The observed values of precision and recall for our approaches are maximal over each dataset but *DBLP*, where XPCMC performs better than XPMCS.

We were not able to successfully complete the tests on *DBLP* with *SGrace* and *XRep*. This is indicated by means of symbol—in the corresponding entries of Table 1.

As far as clustering effectiveness over the *Synth2* collection is concerned, notice that *XProj* also achieves the same maximum precision and recall as our approach. However, it is worth to underline that the performance of the devised approach is measured over 3000 XML documents, whereas the performances of *XProj* was obtained in [3] over a much smaller *Synth2* collection, consisting of 300 XML documents. Additionally, the sensitivity analysis of *XProj* (Fig. 4 and Table 4 in [3]) reveals that an improper setting of important input parameters, such as the sequence length as well as the minimum and maximum support-thresholds, may lower precision over the reduced *Synth2* collection down to nearly 0.92 %.

Table 1 also confirms that XPMCS is generally capable to autonomously discover the actual number of clusters in the XML data. Cluster number is instead a problematic input parameter of the chosen competitors.

Yet, efficiency is a strong point of XPMCS and XPCMC.

By looking at Table 1, one can notice that the overall running time taken by the devised approaches is up to three orders of magnitude less than the running time of the chosen competitors. We emphasize that the running times of our approaches actually include the time for the prior enumeration of the individual substructures in the underlying collection of XML documents.

6 Conclusions and Further Work

Focusing on only one form of structural components in XML partitioning may not suffice to distil structurally-homogeneous clusters. To overcome such a limitation, two parameter-free structure-oriented approaches to XML document partitioning were proposed. Both allow to consider multiple forms of structural components to separate structurally-homogeneous clusters of XML documents.

Experimental results obtained over both real and synthetic XML data proved that the effectiveness of the devised approaches is at least on a par and even superior with respect to the effectiveness of state-of-the-art competitors. Additionally, the empirical evidence also revealed that our approaches outperform such competitors in terms of time efficiency.

We planned to extend the framework with the analysis of content features. This is useful for effective clustering when XML documents share an undifferentiated structure. Various algorithms are in principle eligible for developing such an extended approach, such as co-clustering and nearest-neighbor clustering algorithms [30, 31]. In particular, nearest-neighbor clustering was found highly effective in data de-duplication and can be expedited through hash-based indexing for fast (approximated) similarity search [31].

References

1. S. Abiteboul, P. Buneman, D. Suciu, *Data on the Web: From Relations to Semistructured Data and XML* (Morgan Kaufmann, 2000)
2. E. Wilde, R. Glushko, Xml fever. Commun. ACM **51**(7), 40–46 (2008)
3. C.C. Aggarwal et al., XProJ: a framework for projected structural clustering of XML documents, in *Proceedings of SIGKDD International Conference on Knowledge Discovery and Data Mining (KDD)* (2007), pp. 46–55
4. R.A. Baeza-Yates, N. Fuhr, Y. Andamaarek, Special issue on XML retrieval. ACM Trans. Inf. Syst. **24**(4) (2006)
5. L. Denoyer, P. Gallinari, Overview of the INEX 2008 XML mining track, in *Advances in Focused Retrieval* (2009), pp. 401–411

6. T. Asai et al., Efficient substructure discovery from large semi-structured data, in *Proceedings of Siam Conference on Data Mining (SDM)* (2002), pp. 158–174
7. K. Wang, H. Liu, Discovering typical structures of documents: a road map approach, in *Proceedings of ACM SIGIR Conference on Research and Development in Information Retrieval (SIGIR)* (1998), pp. 146–154
8. A. Algergawy, M. Mesiti, R. Nayak, G. Saake, XML data clustering: an overview. ACM Comput. Surv. **43**(4), 25:1–25:41 (2011)
9. L. Denoyer, P. Gallinari, Report on the XML mining track at INEX 2007: categorization and clustering of XML documents. ACM SIGIR Forum **42**(1), 22–28 (2008)
10. G. Demartini et al., Report on the XML mining track at INEX 2008: categorization and clustering of XML documents. ACM SIGIR Forum **43**(1), 17–36 (2009)
11. R. Nayak et al., Overview of the INEX 2009 XML mining track: clustering and classification of XML documents, in *Focused Retrieval and Evaluation* (2010), pp. 366–378
12. M.N. Garofalakis et al., XTRACT: a system for extracting document type descriptors from XML documents, in *Proceedings of International Conference on Management of Data (SIGMOD)* (2000), pp. 165–176
13. S. Nestorov, S. Abiteboul, R. Motwani, Extracting schema from semistructured data, in *Proceedings of ACM SIGMOD International Conference on Management of Data (SIGMOD)* (1998), pp. 295–306
14. S. Bergamaschi, S. Castano, M. Vincini, Semantic integration of semistructured and structured data sources. SIGMOD Record **28**(1), 54–59 (1999)
15. G. Costa, G. Manco, R. Ortale, A. Tagarelli, A tree-based approach to clustering XML documents by structure, in *Proceedings of International Conference on Principles and Practice of Knowledge Discovery in Databases (PKDD)* (2004), pp. 137–148
16. S. Joshi, N. Agrawal, R. Krishnapuram, S. Negi, A bag of paths model for measuring structural similarity in web documents, in *ACM SIGKDD International Conference on Knowledge Discovery and Data Mining* (2003), pp. 577–582
17. G. Costa, R. Ortale, E. Ritacco, Effective XML classification using content and structural information via rule learning, in *IEEE International Conference on Tools with Artificial Intelligence* (2011), pp. 102–109
18. G. Costa, R. Ortale, On effective XML clustering by path commonality: an efficient and scalable algorithm, in *IEEE International Conference on Tools with Artificial Intelligence* (2012), pp. 389–396
19. G. Costa, R. Ortale, E. Ritacco, X-class: associative classification of XML documents by structure. ACM Trans. Inf. Syst. **31**(1), 3:1–3:40 (2013)
20. T. Dalamagas, T. Cheng, K.-J. Winkel, T.K. Sellis, A methodology for clustering XML documents by structure. Inf. Syst. **31**(3), 187–228 (2006)
21. F.D. Francesca, G. Gordano, R. Ortale, A. Tagarelli, Distance-based clustering of XML documents, in *International ECML/PKDD Workshop on Mining Graphs, Trees and Sequences* (2003), pp. 75–78
22. M.J. Zaki, C.C. Aggarwal, Xrules: an effective structural classifier for XML data, in *Proceedings of SIGKDD International Conference on Knowledge Discovery and Data Mining (KDD)*, (2003) pp. 316–325
23. W. Lian, D.W.-L. Cheung, N. Mamoulis, S.-M. Yiu, An efficient and scalable algorithm for clustering XML documents by structure. IEEE Trans. Knowl. Data Eng. **16**(1), 82–96 (2004)
24. G. Costa, G. Manco, R. Ortale, E. Ritacco, Hierarchical clustering of XML documents focused on structural components. Data Knowl. Eng. **84**, 26–46 (2013)
25. G. Costa, R. Ortale, Structure-oriented clustering of XML documents: a transactional approach, in *IEEE International Conference on Intelligent Systems* (2012), pp. 188–193
26. M.J. Zaki, Efficiently mining frequent trees in a forest: algorithms and applications. IEEE Trans. Knowl. Data Eng. **17**(8), 1021–1035 (2005)
27. E. Cesario, G. Manco, R. Ortale, Top-down parameter-free clustering of high-dimensional categorical data. IEEE Trans. Knowl. Data Eng. **19**(12), 1607–1624 (2007)

28. T. Li, M. Ogihara, S. Ma, On combining multiple clusterings: an overview and a new perspective. Appl. Intell. **33**(2), 207–219 (2010)
29. R. Baeza-Yates, B. Ribeiro-Neto, *Modern Information Retrieval* (Addison-Wesley, Boston, 1999)
30. G. Costa, G. Manco, R. Ortale, A hierarchical model-based approach to co-clustering high-dimensional data, in *Proceedings of ACM Symposium on Applied Computing* (2008), pp. 886–890
31. G. Costa, G. Manco, R. Ortale, An incremental clustering scheme for data de-duplication. Data Min. Knowl. Disc. **20**(1), 152–187 (2010)

Security Applications Using Puzzle and Other Intelligent Methods

Vladimir Jotsov and Vassil Sgurev

Abstract New types of constraints are considered. Reasons are described that lead to applications of intelligent, logic-based methods aiming at reduction of risk factors to ATMs. Special attention is paid to applications of Puzzle method in ATMs. To make a more independently functioning ATM, the proposed methods should be applied to data/knowledge/metaknowledge elicitation, knowledge refinement, analysis of different logical connections aiming at information checks.

1 Introduction

Financial losses due to cyber-criminality constantly increase worldwide and today they exceed one trillion US dollars which equals to an 'imaginary' eighth in magnitude worldwide economics. Here we shall show that a great deal of the accumulated problems is solvable no only via highly intelligent technologies but also applying elementary logical operations at the technical level. Also, introducing more and more sophisticated devices is sometimes arguable because they are vulnerable for more unexpected threats; they are more expensive and they require permanent high cost support.

Accumulated problems show that in cases of operations with cash machines or in on-line payments still there are elementary logistic problems. For example, receiving money from an ATM and if at the same time with the data from the same card money is drawn on-line or from other money devices, then all transactions must be locked immediately. Reports from some US banks show [1] that intervals between different transactions are often compared manually as an assumption; still there is no data about systems for automatic monitoring and (re)action. For this

V. Jotsov (✉) · V. Sgurev
State University for Library Studies and IT, Institute of Information
and Communication Technologies, Bulgarian Academy of Sciences,
P.O. Box 161, Sofia 1113, Bulgaria
e-mail: jotsov@ieee.org

© Springer International Publishing Switzerland 2016
M. Hadjiski et al. (eds.), *Novel Applications of Intelligent Systems*,
Studies in Computational Intelligence 586, DOI 10.1007/978-3-319-14194-7_10

reason it is suitable to organize interbank systems for monitoring with the least possible information, to prevent using confidential data: e.g. not the *amount* but the *time* and the *place* of the transaction.

Reducing losses in US banks is realized for example by experiments with checks of biometric data in ATM devices [2]. Checks in Europe of voice data and also of clients' images led to identifications of terrorists. It is expected soon in Sberbank ATM devices in Russia to be introduced fingerprint checks, checks of voice data [3]; also if the user responds to the inquiries from the machine nervously or if some of the questions provoke negative emotions and stress then the machine will fix these states and report the problem(s).

On the background of the cited changes, this paper presents some software changes which practically do not require not only significant hardware changes but also they avoid significant changes in the usual client's operations with the cash machines.

A novel Puzzle method is described in our book [4]. It enables fixing hidden correlations, causal links and dependencies via using a system of heterogeneous quantitative and qualitative constraints and solving logical problems of relating unknown knowledge to the known one. The Puzzle method must be distinguished from just syntactic procedures for filling crossword puzzles using some definite set of words [5, 6]. Its applications differ substantially from other security applications. The quoted here security applications deal mainly with syntactic procedures and cryptographic problems also named Puzzle [7–9]. On the other hand the considered Puzzle method also belongs to the group of constraint-satisfaction methods [10–16] but is data/web mining oriented and semantic by nature. Its applications are discussed in detail in Sect. 2. Ontologies are introduced in Sect. 3 to raise effectiveness of the applications. Ontologies are increasingly used for the needs of security [17–23]; they are most often used for relations 'is-a' and 'has-a' as in the case with Protégé [24]. Dynamic multimedia and sound ontologies may be also used most of all for training but there is a profound difference between ontologies used for cognitive agents and agents of type semantic reasoner. The present report is oriented just at using reasoners. Few application aspects are presented in Sect. 4.

2 Experiments with Puzzle Methods

Let us assume that the reader resolves a problem with a complex sentence of 400 letters with vague for the reader explanations, for example: 'elaborate a good practice for the crisis management'. Let the unknown sentence be horizontally located. The reader can't resolve the problem in an arbitrary manner, because the number of combinations is increased exponentially. Hence the traditional syntactic approach on the base of all the possible word combinations is inapplicable in case. Now it is convenient to facilitate the solution by linking the well-known information with the unknown one from the same model. Let's say, the reader tries to find easy recognizable vertical words that he is conscious about like the largest US banks—Citibank, Bank of America and a few more. If he fixes the length of the unknown word as it is in the case of the classical crossword puzzle then the searched set furthermore gets narrower; but even in the worst case the presented below Puzzle method will still work. The more the known cross-points are, the easier is the solution of the desired horizontal sentence. The approach for the Puzzle is even easier. Here both the easy-to-solve meanings and the difficult ones are from the same field, therefore additional clues exist to accelerate the process of finding the final solution.

The difference of the Puzzle method from the usual crosswords is in its semantic (logical-oriented) nature, in the logical processing of interconnected constructions, in the usage of nonlinear and/or ill-defined (ontology based etc.) constraints, and in high-dimensional space applications where the length of the crossing information lines may be unknown.

Puzzle solutions use a set of mutually connected pieces of knowledge denoted as $K_i \in K$ where K is the set of all accumulated knowledge. Also $K_i \in K^*$ where $K^* \in K$ is a set of knowledge bounded by different logical connections: in classical case they are implication, negation, disjunction and conjunction. As shown below, different nonclassical analogs may be effectively used.

I. The meaning of the classical implication "A is connected to B" is following. First of all, from "A is true" it follows that "B is true" or "from B it follows that A".

II. Different nonclassical causal relations are frequently used in practice of contemporary intelligent systems. For example, "IF E is true THEN C *defeats* D" where E is one of the causal relations in form of a fact where any relations between any literals in D may be deleted or changed by an arbitrary set of other literals or the whole rule may be modified or the defeated rule may be changed by a different rule [4].

III. M has some connection to N if their literals are included in one formula like, for example, 'M or N…', or are one knowledge part e.g. ontology etc., in other words: M bounds N.

In situation I classical inference schemes lead to formation of colorraries A, B, C …, etc., formed after deductive applications. In such a way a chain line of mutually connected knowledge pieces is formed as depicted in Fig. 2, line E below. On the other hand, same constraint line may represent just one fact like 'zone of intrusion activities'.

In situations II or III nonclassical relations should be used as described below Fig. 2.

Let G be a goal that must be solved. Figure 1 shows the decomposition of the goal into two types of goals: G_2 is inferred in the logical manner and G_1 is explored in the area that is bounded in the left part of Fig. 1.

Let the goal G_1 is indeterminate or it is defined in a fuzzy way. Then the introduced algorithm is defined in the following way.

Fig. 1 Decomposition of the goal to different types of subgoals

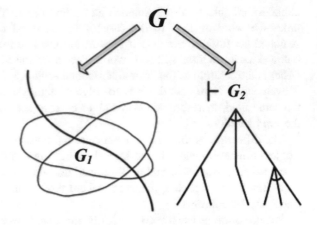

Fig. 2 Spatial area for resolution of the goal G_1

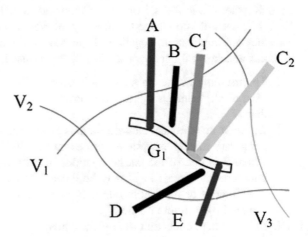

$$K_i \in K, \quad i = 1, 2, \ldots, n : G_1 \cap K_i \neq \emptyset; \tag{1}$$

$$L_j \in L, \quad j = 1, 2, \ldots, m : G_1 \cap L_j \neq \emptyset; \tag{2}$$

$$S = (G_1 \cap K_1), \quad T = (G_1 \cap K_n); \quad S \neq T; \quad x_1, y_1, z_1 \in S; \tag{3}$$

$$\frac{x - x_1}{x_2 - x_1} = \frac{y - y_1}{y_2 - y_1} = \frac{z - z_1}{z_2 - z_1}; \tag{4}$$

where x_1, y_1, z_1 and x_2, y_2, z_2 are the coordinates of the respective boundary points from S and T from the set K whilst x, y and z are the coordinates of the points from the slice that tethers the explored area. In this way—by two sticking points—the goal search is restricted from an infinite space to a slice in the space. The introduced method is realized in an iterative manner: the goal place from (1) is replaced by from the previous iteration and so on. Linear constraints are presented in (4) but with the same success it is possible to use in the method also non-linear constraints and informalizable/fuzzy ones in the form of ontologies.

Figure 2 illustrates an example with standard linear and non-linear constraints consisting of three elements where K is an own subset of G so the elements of K are called 'crossword constraints'. The third type of constraints is contained in the set of bounding constraints and in our case it contains two elements B and D. The bounding constraints from L do not include elements from the goal G_1 but they are neighbors of it, they show proximity to the goal and they may present a series of weakly formalized or 'informal' causal relations and other forms of knowledge representation. The example illustrates the benefit from the introduction of any elements of L or from the spatial constraints even when it is conspicuous that the direction of G_1 most often does not predetermine the integral decision and that the elements D, E and V_i decrease the number of the possible alternatives.

The V_i constraints of linear/nonlinear form are the classic case borrowed from constraint satisfaction methods. The B–D binding constraints doesn't intercept the unknown goal but are located close to it, they reveal a 'neighboring' area around the target. The A–C–E constraints are named the crossword constraints because their interception with G_1 gives a part of the searched goal where C_j form larger interconnected areas in G_1. The usage of the considered set of C_j-s is much more effective than of the classical set of V_i-s.

Through binding constraints it is convenient to implement the many non-classical causal relations of the type "A is linked to B but the connection between them is not implicative". The same could be presented through heuristics, which is not recommendable. It must be pointed out that through this type of constraints the location of the searched solutions is fixed in a way that is best combined with fuzzy methods.

Crossword constraints offer new ways of assessment (outlook) for the searched unknown solutions on the basis of the accumulated so far knowledge.

2.1 On the Usage of Different Binding Constraints

The following section discusses the introduction of three types of binding constraints.

The knowledge used in the Puzzle method can be presented as parts of information (atoms), linked by different relations. Usually these relations have been obtained by logical processing of information like structuring, extracting meaning from information blocks or other processing. In this terminology, one rule can be presented in the following way:

The conjunctions of antecedents are $A_1, A_2 \ldots. A_z$.
The conclusion/consequent is marked as B.

Let all z number of conjunctions are proved to be true/confirmed, then B is true, whereby the goal/problem of checking whether B is true has been solved (see Fig. 3).

When significance has not been pointed out, as in the case shown in Fig. 3, then the significance of the conjunctions are considered equal (1/z). The bows show that each conjunctive has its individual significance in proving the conclusion: a number between 0 and 1. But in the common case, some conjunctives are of great significance while others are less important, depending on the situation. Part of these invisible links between the atoms of the type conjunctive-conclusion or other parts of knowledge can be torn in different conditions, for example when there is additional information and during a process called defeat, which can change the truthfulness (the truth value) of each atom, reduce its *significance* in proving the conclusion/goal to zero or change/*shift* the whole rule completely, for example, by replacing it with another one.

The defeasible process is started by special forms of presenting knowledge called exceptions/*exclusions* to rules (see Fig. 2), knowledge of the type E (C, A_p), where a prerequisite for the defeat is the argument C of the exception, which must be true in order to start this process of defeat. Therefore, both arguments of the exception enter a causal relation, which is not implicative. It is a kind of non-classic and many times informal causal relation. In life we use so many similar relations and most of them are difficult to formalize because they are not included in the classical mathematical and formal logic (Fig. 4).

The accumulation of such parts of knowledge/atoms of knowledge, compound with different classical and non-classical relations allows us to use new opportunities for reaching the set of goals.

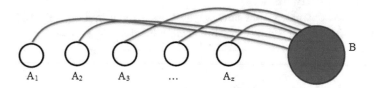

Fig. 3 Relations inside a rule

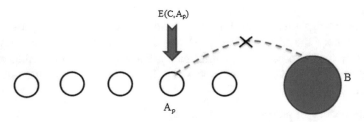

Fig. 4 How to defeat implicative connections

Fig. 5 The binding process

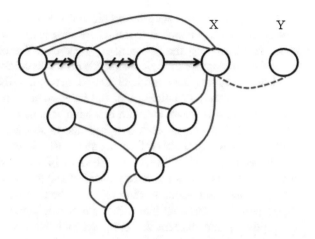

For example, let's have goal X proved via classical or non-classical means, as shown in Fig. 5 and let X and Y have an unidentified causal relation, for example X defeats Y; or for example, let X and Y be statistically linked variables. In this case, the fact that there is a large volume of information linked to X, as shown in the figure, leads to imposing informal constraints over the choice of the condition of Y, regardless of the fact that in a classical logical (formal) sense, X and Y are not linked. Informally, proving X leads to the confirmation of the goal Y. In other analogous situations, solving X does not lead to proving Y. However, they are linked through the following non-classical causal relation: proving X shows that we are close to the solution of Y, the more rules and facts prove the truthfulness of X, the higher the confidence/sureness/certainty or belief in the hypothesis that Y is true. The described process will be called binding, as interpreted in Fig. 5.

Very often using methods like fuzzy logic helps binding, for example, in situations where indefinite notions are used or notions that change their meaning depending on the situation and the context.

2.2 On the Usage of Fuzzy and Ontology Applications

Let us examine an example when the iterations do not lead to a final answer. Let G_1 show 'an industrial application of fuzzy logic'. Due to an incompleteness of the domain it is not possible to reach an exact answer and let the expert receives a question: 'where in cash machines are applied fuzzy measures?' In case of failure the suggested method still can try discover new constraints in the investigated model so as to lower the alternatives and choose the best solution.

Our recent research shows that usage of ontologies with fuzzy perceptions or measures could avoid a set of hard-to-formulate and operate boundaries and/or binding elements.

The Puzzle method security applications consist of modeling and exploring a system of heterogeneous constraints that logically lead to narrowing the search space or directly to the solution; they increase the effectiveness of the applied algorithms via applying logical analysis of the situation including the user actions. Modeling includes also artificial neural network (ANN) training of ATMs for operative standard situations, e.g. the consequence 'read the information on the screen, the card is drawn out with one hand, it is placed in the device, after that the PIN code in introduced, etc'. It is possible to include ontologies in the model to explain the sense of matters. It seems that most *effective* is the usage of the relations HOW, WHY, IMPORTANT described in [4]. Applying ontologies is a costly process but it increases the effectiveness of using devices that can not only analyze facts but also the reasons leading to them, the meaning of actions, etc. It has been experimentally found out that using non-classic logics, in the first place analogies, descriptive, paraconsistent and temporal logics, universalizes the applications. For pity, at present they are not included in one single research due to the high level of algorithmic complexity of sophisticated applications which is usually lowered by using heuristics. The latter substantially narrows the learning capabilities and applicability of elaborations, hence we don't use such strategies. New evolutionary applications are experimented that could allow the construct of more complicated elaborations.

The so presented model is defined as a set, where L is a machine training set, O are ontologies, T are the current user's activities data and U is the accumulated information from the users. In the present version U is the knowledge accumulated at one place but it is clear that communicating between separate devices in a united MAS system will improve the applications. During the operation process T are compared to L in an iterative manner using U and with or without using O. After each iteration a part or all knowledge from T is copied to U. Analyzing T is activated in cases when there is a significant number of discrepancies with L or by a command of the system administrator.

2.3 Example

Discrepancies between T and L are compared using different metrics and the well-known Occam's Razor principle. In our case it is realized also by the principle of convenience for the user. Here the convenience is related to the notion of conflict where anti-conflict means convenience, accordance, comfort. The common idea here is that no one undertakes actions that are unsuitable for him/her without some definite reason. Let's say, if the card holder does not fix the card exclusively by one of his/her hands but moves also the other hand at the same time then this is inconvenient according to the model in the well trained system. The user definitely knows that this activity will not speed up the procedure that way but the cardholder may retard it instead, e.g. if the card isn't properly fixed in the machine. Here the question is: *why* the card holder is doing so; the answer is related to using different types of knowledge. This may be due to nervousness, hurry, also it may be because some intruder uses one of his hands for masking gestures whilst he puts a device inside the machine, sticks trackers onto the keys, etc. Hence in this situation, in spite if its bunch of meanings, an exhaustive analysis of user's actions is required. The analysis doesn't significantly deviate from the standard situation: asking additional questions and analyzing the replies, at least one camera video rewind aiming to analyze the user activities/correspondence to other persons where the goal is to find a well-known intruder. But locking the card and even a delay in the card processing will offend the reputation of financial and maintenance institutions. In this example such an analysis is activated while in the ordinary situations this isn't the case. Let's discrepancy between T and L is revealed because the human is clicking on the keys w/o properly inserted card. Is he a mad man or simply cracker? No matter of the reason, alarm should be activated and let the competent staff decide what is the matter. More information of such cases is accumulated in Fig. 7 and descriptions below it.

The present research explores also the possibility to reveal new rules applying the same system of three types of constraints (Fig. 2). The idea is to constrain and use binding constraints to fix up the relation 'the consequence from A is B' which is not obligatory a rule using classic implications. In the presented domain the problem is simplified to 'the consequence from A is an intrusion' which increases the effectiveness of the applications. Binding constraints represent the experience of the system but they are not obligatory knowledge of any heuristic type.

2.4 Example

A kid plays with a toy-helicopter near to an ATM. Let a microphone is installed in the ATM for communication with clients. At every accidental approach of the toy close to the money device the communication is locked by the noise. Now we have neither of the standard linear or non-linear constraints as well as the crossword constraints. Just the binding constraints have been used and the lesson is that instead of forbidding children to play near to ATMs, a rule must be added forbidding mechanical devices appearance inside the close perimeter of the ATM and for automatically activation of an alarm provoked by the occurrence and an increase of mechanical noise which is easy to be trained by classical ANN machine learning. Still it is hard to believe that this will be included in the standard procedures for ATM learning, so we consider this example as a reaction to unknown threat. The reasons for all unknown threats can't be enumerated and learned in advance but they cause a limited set of damages so the result in the rule can be predicted and automatically processed. When the result is 'mic locked', 'cam blinded', 'dispensing mechanism doesn't work properly', 'PIN pad doesn't react on touch' and in few other situations the reaction of the ATM system should be known let's say, 'lock everything and alarm'. After result B occurred, the reason concerning the premise A may be resolved by a security team gathering at place. On the other hand, the machine reaction is incomparably faster in case, and the above example shows up how automatically to link the known fact (B) to the unknown to the system knowledge: reason A. Here the Puzzle method provides a number of new and original solutions. Let somebody or a group of people use a laser pointer close to an ATM and accidentally or on purpose the ray penetrates in the camera. The reason (A) may be unknown, i.e. the system lacks experience in case, it may not be incorporated in the training course but the consequence is evident: locking the video-camera operation mode. This and other similar cases like locking any other ATM function leads to an automatic locking the device operation and also to an alert of the team on duty. As a result from this research we recommend that near to any ATM there should be at least two cameras: one of them may be cheap stationary cam, and the other must rotate to monitor fixed targets.

2.5   Example

The case when the intrusion is proven. If the card holder uses 'the white plastics' or a card with a non-standard font, a non-existent name/number or with any other false data then it comes a proven intrusion, the card must be hold and the intruder must be invited for a conversation with the bank officers to clarify the case in case the discrepancy between T and L goes above the established threshold. Hence we recommend installation of cheap color/pattern recognition systems at every device.

On the other hand, if the card holder confuses the PIN code then there is no intrusion yet. But if there are several unsuccessful attempts using same card to penetrate into the bank system from the web sites or from different ATMs then there is a suspicion for an intrusion which must be clarified. The presence not of just one but of three or even more errors also characterizes card operations as suspicious, in many cases a clear intrusion is ascertained. The accumulation of errors must lead not to a linear but to an exponential level of increasing attention. For example if there are disturbances in machine operation and also if there a PIN code error appears then all must be locked. Figure 6 shows a stylized graph of user's actions. These actions together mark a closed area with the intrusion encompassing the zone between points M and N. Sometimes we are unable to fix the type of the relation between the two points but we do fix the cluster of the intrusion between M and N; in many cases this is enough to alarm the situation. This cluster of granule reveals different types of causal relations including the rule A → B.

In this case we cannot be sure with 100 % confidence that it comes an ATM intrusion but there was an alarm and the situation will be clarified after checking whether the person is a regular cardholder and after the person answers a series of questions asked via the ATM or in the office of the bank.

Lots of intrusions consist not just of one single but a series of harmless at first glance actions.

Fig. 6 Constraint area for the
task type M causes N

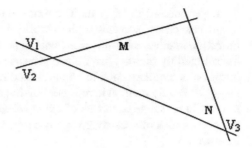

Figure 7 introduces a graph of the ATM reaction to a combination of different users' actions where

- Set 1 is 'user group interference', i.e. a group of interconnected persons including the cardholder is revealed (sf. above);
- Set 2 means 'malfunctioning hardware';
- Set 3 is 'error action consequence';
- Set 4 means 'repeating errors';
- Set 5 is 'touching when and where shouldn't';
- Set 6 is 'non-comfort actions', i.e. somebody is walking backwards.

All crossings with sets 1 and 2 lead to automatically locking the system (marked in dark color) and alert the security team. In some situations the result is due to negligence, in others—due to personal problems, in third it is done deliberately; for the system this does not matter, it is the result that makes sense. Subsequently the experts will have time to distinguish between the situations and they will make appropriate conclusions, the system helps them with its timely alarm.

The brighter colored area in Fig. 7 excludes locking everything but the described above profound analysis of user's actions. For example if the cardholder data is high growth and he/she is less than a meter in case then control is transferred to the worst option marked by the dark area in Fig. 7. White areas in Fig. 7 do not involve additional processing but in our projects this data is also stored thus accumulating

Fig. 7 System response on
cardholder activities

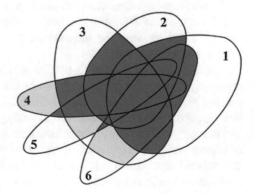

the experience of the system. The monotone dark region in Fig. 7 represents different risk cases: sometimes the security team must immediately apprehend the intruders, in other cases it is sufficient the person to come in suitable for him time in the selected by him/her bank office. The separation of these types of reaction to an intrusion is recommended to be performed manually because the common goal of using the Puzzle method here is not making automatic programs for everything but to react in real time to threats of unknown type and therefore improve and expand the possibilities of security groups to properly orient in large volumes of incoming information.

It can be statistically proven that many people take off their gloves before using an ATM. It is clear why this is so: evidently the ATM is outside where sometimes it is cold: the temperature is accessible from regional Web sites. It is uncomfortable to operate with the card with gloves on the hands. The machine can make similar inferences about the sense of such actions after a deep knowledge processing which must follow modeling the domain combining manual and machine information input. But even without understanding the idea of user's actions ATM with the recommended card/user pattern recognition software will conclude that many people take off their gloves before transactions, more people operate without gloves and only a very tiny group of cardholders keep or put on gloves before they introduce the PIN codes. As it is inconvenient to input data this way this group of people acts in a very exotic way and therefore it is right that the ATM should record user's actions in separate files and then security staff will be asked about such actions. What the staff will answer? User's actions in the case aren't intrusion: certain people prefer gloves for clean hands, etc. but generally acting with gloves on the hands may be a feature for hiding fingerprints. It is necessary to collect more data about this group (biometric, etc.), the results must be kept at a special place; if there is a suspicion about an intrusion then this information must be forwarded to experts investigating the case. Such actions schematically depicted by lines B and D in Fig. 2 do not determine an intrusion but they are related to a part of the intrusions—hiding information about oneself, wearing a hood, hiding the face in front of an ATM, wearing gloves is sometimes suspicious. Intrusion depends on the situation: if other hidden players from the group are revealed then the area goes from light to dark areas in Fig. 7. The presented type of constraints is the binding constraints.

Constraints can transform from one type into another depending on the situation and the context of the intrusion. Let us reveal on examples how the binding constraints transform into the crossword/classic linear/classic nonlinear (sometimes fuzzy) constraints.

The third type of constraints used here is the crossword constraint. Any of them separately shows just a part of the intrusion type M–N. For example if someone touches an ATM at different places before fixing the card still this is not an intrusion but a great deal from the intrusions (mounting skimmers, etc.) are accompanied just by such actions. Therefore in such cases the ATM must apply the whole existing set of checks which is inapplicable in the standard case due to client's delays, etc. But the pointed out fact is not a prerequisite for an intrusion: e.g. the person has problems with the eyes, with the coordination, with drugs/alcohol so the behavior is of this type. Here we may apply an analogy with a puzzle: if one guesses a long word letter by letter then it is quite likely to go amiss. If the client also does not finish the transaction and behaves in a non-standard way, goes around, bides, etc. then surely it comes an alleged intrusion. By analogy with the crossword constraint in this case we have more and more intersections with the area M–N so the intrusion becomes more and more probable.

Like in the first case of the classic constraint satisfaction, in the third case the intrusion is related to a system of knowledge formulated as constraints but in the case with situation 1 the area contains the proven intrusion whilst in the case with situation 3 more and more (puzzle) fragments of the intrusion are revealed that are logically related in a whole. The combination of acquired knowledge for the situations from 1 to 3 gives even a better picture of the intrusion or of its negation, i.e. of the suspicious user's actions that lead to no intrusion.

Essentially in the presented examples the ATM control system does not seek an intrusion, instead it establishes series of dependencies of the type 'people wear gloves because it is cold' on the basis of temperature data 'today it is cold', 'people wear sunglasses because it is sunny', etc. In this way via applying knowledge for constraining the set of possible meanings rules are established 'from M it follows N' (Fig. 6). Intrusions are fixed otherwise, e.g. using heuristics 'rare phenomena must be studied in detail' which means obtaining biometric data if any, whenever possible asking questions, monitoring gestures and the reaction of the client, delaying card transactions without explaining the reasons, accumulated user knowledge base processing, etc. So if during the late hours the client wears sunglasses or if he is near the ATM much too long—20–30 min while other clients are through with the operation for about 2–5 min then clients' actions are classified as suspicious and if he/she operates with a card in the ATM then the data are copied together with the video fragment to a certain place 'warning: this activity should be analyzed'.

Crossword constraints are illustrated by next example 'group with flags'. Let a group of intruders appears in front of an ATM one by one, the first of them wears a flag and the flag is placed too close the camera. This is not an intrusion, it is a harmless action, but it enables the intruders to act disguised. It is impossible to record all similar suspicious actions. The ATM must be prepared in advance: if light sharply decreases or if the scenario all of a sudden lost its dynamics then the ATM operation is locked and an alert is initiated. If ATM operation is locked at the presence of people around then the mode also changes to 'danger': the client may lock the device on purpose, e.g. spreading the adhesive or liquid nitrogen on the card, or make other unpredicted activities. Intruder's actions are unpredictable but elementary learned experience together with the Puzzle method applications allow relate unknown facts with known ones: locking the device or its security system. The purpose here is not to reveal all possible dangerous actions but to add secure tools for self-analysis. In the example with the flags after closing the cam picture any locking of the device or of other non-standard modes (PIN error, etc.) should lead to a signal for intrusion; the picture is clarified via collecting together different fragments from the situation. On the other hand, the reaction to pointing a laser ray right at the cameras must be directly triggering an alarm.

The more sophisticated the ATM, the more unexpected ways for intrusions may happen. If the case is with a camera using methods for image processing or a camera directly transmitting the picture to the security administrator then it is possible to play out a scene to diverse via sharp gestures, unexpected actions like for example suddenly dropping a wig, etc. In this case, if there is just one camera, actions are classified as *binding constraints* related to intrusions; therefore the attention must be concentrated over ATM-functioning and its security.

3 Helpful Ontologies

The system can understand the sense in client's actions in different situations if ontologies are introduced in it, e.g. such made in Protégé [24]. Some difficult to determine phrases, e.g. suspicious actions, actions to distract attention, etc. are best expressed via ontologies (Fig. 8), e.g. ontology O is 'suspicious activities near to the ATM' and ontology O_1 is 'increasing incidence of on-line thefts of identity'.

This is the way to search necessary goals not only in linear or non-linear constrained areas in Fig. 2 but also by crossing two or more previously introduced ontologies (Fig. 8). For brevity we shall discuss just the most popular ontologies of the vocabulary type. For example, the first ontology is 'what it sounds like the noise of mechanical devices' and the second ontology is 'how the sound grew'. The crossing of both ontologies contains the solution concerning the purpose: if the noise from a mechanical device increases and it intrudes a forbidden perimeter then ATM activates the alarm. In the described in Fig. 8 case the proximity between the meanings of two ontologies components is estimated not just by a direct congruence but also using different assessments of the concepts contained in the nodes of the graph. For example, the coincidence of fuzzy meaning 'mechanical noise' is based not on a definite frequency of the sound but on the image of its repeatability. Fuzzy granules can be formed in the crossing.

Fig. 8 Search for causes of the intrusion at the intersection of two ontologies

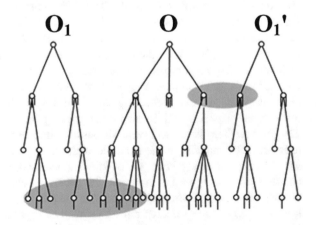

Both ontologies are shown as graphs and they represent the sense of things of the type 'two objects are linked by the relation R', e.g. $object_2$ is a feature of $object_1$ or $object_1$ is a predecessor of $object_2$. In Fig. 8 with curved areas are denoted crossings between the two ontologies where the searched solution is located. If both ontologies concern sounds, multimedia or some other type of unimaginable via graphs information then their crossing is found otherwise but this does not significantly change the proposed search algorithm. Using operations with ontologies instead of classic set operations gives the advantage of faster information processing, natural usage of rather fuzzy boundaries that are hard for formalization otherwise, and so on. The following situation is presented in Fig. 8: O_1 and O_1' are one and the same ontology. The graph of O_1 is moved related to the graph of O when the situation changes which in turn changes also the sought solution depicted on the graph by the elliptical area. For example, let us analyze a situation almost impossible in most contemporary ATMs. If the system explores the increasing mechanical noise but the noise is used for diversion and during this time the intruder uses the fact that unsolicited note will be harvested by ATM, the cardholder intruder sets between them a tracker and waits until it is stowed in the machine. The case with the sought intrusion is not in the left but in the right elliptical zone in Fig. 8. In this case the system must include the alarm at least due to situation 1 from Fig. 7: detected group attack.

Figure 9 presents knowledge that if some mechanical device comes closer to the ATM or if a laser is pointed at it or if the case is with a shock impact then the alarm system must be immediately activated and the operation of the ATM must be temporarily locked. As a result of the usage of an ontology like the depicted on Fig. 9 we obtained the results from Fig. 7. The case uses high-level terms. How can be detected these situations? This depends on the equipment of the respective bank. For example if there are sound recorders installed then ATM is trained to detect mechanical devices also by their noise. When such device approaches the ATM at a distance of several meters then the mode changes to 'caution'. Lasers are detected by their rays before the rays are directed to the camera. Shock impacts are also most easily detectable by the noise, etc. Figure 10 depicts the fact that the attack includes the ATM as well, its video-, sound- and eventually active protection, if any.

Fig. 9 Some instant-alarm situations

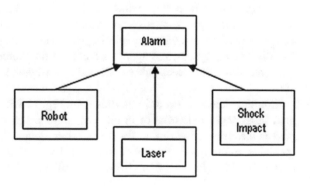

Fig. 10 Ontology of essential
ATM attack components

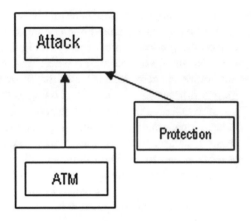

Creating and changing ontologies by the Puzzle method is a subject to research in our university. Both ontologies on Figs. 9 and 10 contain incomplete knowledge and should be gradually improved. Our goal was to show that even an incomplete knowledge can lead to significant results when using the Puzzle method.

The future for ATM protections is multi-agent ATM systems with information (experience) transfer via ontologies and also with the presented above logical control for intrusion protection.

4 Application Results and Examples

The experimental system related to the presented project includes an adaptation of several popular semantic reasoners via adding Java code to change the analysis and semantic conflict resolutions and also input of different combinations of elements guided by the Puzzle method.

The presented system source codes are written in Java. All the ongoing research is done in Java because the multi-agent FIPA-based software has been used. An application example has been shown below.

A fragment of Java class Estimator is shown in Fig. 11. An Estimator receives messages from agents capable to execute a chosen goal and is calculating their competence. The first agent or set of agents with enough high competence will get the job. The competence of agents is a constantly changing value depending on job type and number of sweepstakes collected by an agent, let's say, set of realized goals, etc.

Conflict situations occur in module from Fig. 11 when the agent can't execute its goal in an acceptable time or in other situations. Let's say, the agent's way to the goal is interrupted by a route for many other agents. The standard way frequently quoted in multi-agent systems is 'step out' for all conflicting agents. Then another routes should be found, otherwise same conflicts will repetitively occur. Meanwhile

```
            //private Message Template mt;
        //the template to receive replies

        private class Estimator extends CyclicBehaviour{
        private int threshold = 60;
        private char sector = 'A';
        public void action() {
        MessageTemplate mt = MessageTemplate.MatchPerformative(
        ACLMessage.ACCEPT_PROPOSAL);
        ACLMessage msg = myAgent.receive(mt);
        if(msg != null) {
        //PROPOSAL Message received and waiting to be processed
        String title = msg.getContent();
        ACLMessage reply=msg.createReply();
        Integer competence = (Integer) goalSet.remove(goal);
        if(competence >= threshold) {
        reply.setPerformative(ACLMessage.INFORM);
        System.out.println(goal+" in sector "+sector+ " hold by agent
        "+msg.getSender().getName();
        }
        clsc {
        reply.setPerformative(ACLMessage.FAILURE);
        reply.setContent("not-available");
        }
        myAgent.send(reply);
        }
        clsc {
        block();
        }
        }
```

Fig. 11 Estimator Java code example

some goals may be lost. On the other hand, one or more of conflict resolution methods considered above resolve the problem using well known class Agent methods halt(); suspend(); activate(); getState(); move(Location where); and sometimes clone(Location where, java.lang.String newName). Several libraries for class Agent had been used. All of them comply with FIPA standards.

Application examples for other methods are much more complex and can't be described here. They show us that the considered methods aren't so scholastic as it may seem at the first glance. Let's say, Puzzle method depicted in Fig. 2 or in Fig. 6 could be effectively used to find intersections between different agent ontologies an so on. The Puzzle effectively works with the other methods considered in [4], and in first place with the considered above method for identification and resolution of semantic conflicts.

Many of the described procedures rely on the usage of different models/ontologies in addition to the domain knowledge thus the latter are meta-knowledge forms. In knowledge-poor environment the human-machine interactions have a great role, and the metaknowledge helps make the dialog more effective and less boring to the human. The dialog forms are divided in 5 categories from 1 = 'informative' to 5 = 'silent' system. Knowledge and metaknowledge fusion is always documented: where the knowledge comes from, etc. This is our main principle: any knowledge is useful and if the system is well organized, it will help us resolve some difficult situations.

We rely on non-symmetric reply to intruders with a motto surprise and win, on the usage of unknown methods code in combination with well-known methods, and on the high speed of automatic reply in some simple cases e.g. to halt the network connection when the attack is detected. If any part of the program is infected or changed aiming at reverse engineering or other goals, then the system will auto-matically erase itself and in some evident cracking cases a harmful reply will follow. The above presented models of users and environment are rather useful in the case. Therefore different Puzzle method realizations are not named intrusion detection/prevention systems but intelligent security systems because they include some limited automatic reply to illegal activities.

Experimental activities in our faculty are developed by other authors—Ph.D. students and assistant professors; for this reason its detailed description goes beyond the present paper.

5   Conclusion

Typical disadvantages of contemporary ATMs are demonstrated. Based on the Puzzle method a way is proposed for direct ATM control by data/knowledge driven methods with unpredictable reactions depending on the situation and the accumu-lated data/knowledge. The Puzzle method also is used for generalized goal-related solutions, in other words for strategic-level solutions. It allows the learned data/knowledge connection to the unknown, untrained situation to conclude if the intrusion exists. In cases with no positive automatic solution for the problem, experts analyze and estimate how far the decision has been reached; in many cases they can by themselves get things done by hand. Even in this case the usage of the considered method improves the results. Examples show that just by software changes alone it is possible to achieve significantly better results than in contemporary ATM devices.

The main advantage of the considered Puzzle method is the possibility knowledge to be gradually structured and improved using current knowledge model measures. Hence the ATM system can respond to unknown, not learned threats. In this way it becomes data-driven.

As a perspective it is necessary ATMs to be transformed in multi-agent systems (MAS) and they may mutually train one another by information, mainly via ontology transfer. In this case the usage of the comparative analysis will significantly increase the security level of the system as a whole without significant material investments to control ATMs. Major changes can be reached using small to medium size projects.

References

1. Insight. Intelligence. Information. http://www.atmsecurity.com/ to date
2. SANS Institute Reading Room site. http://www.sans.org/reading_room/-whitepapers/authentication/biometric-technology-stomps-identity-theft_126 to date
3. POPSCI. The future now. http://www.popsci.com/technology/-article/2011-06/russian-atm-determines-bankers-identity-and-whether-theyre-telling-truth to date
4. V. Jotsov, Intelligent information security systems. Sofia: Za Bukvite – O Pismeneh Publ. House (2010), 278 p
5. Constraint Satisfaction problems. http://www.cis.temple.edu/~giorgio/cis587/readings/constraints.html to date
6. Crossword Puzzles as Constraint Satisfaction Problems http://www.cs.columbia.edu/~evs/ais/finalprojs/steinthal/ to date
7. S. Dhurandher et al., Puzzle solving-based authentication method for enhanced security in spins and its performance evaluation, in *Proceedings of the 14th Communications and Networking Symposium*, San Diego, CA, USA (2011), pp. 5–10
8. S. Kak, On the method of puzzles for key distribution. Int. J. Parallel Prog. **13**(2), 103–109 (1984)
9. M. Hellman, An overview of public key cryptography. IEEE Communications Magazine 50th Anniversary Issue: Landmark 10 Papers, May 2002, pp. 42–49
10. Cassowary http://www.cassowary.net to date
11. Java constraint solver (JaCoP) http://jacop.osolpro.com/ to date
12. P. Van Hentenryck, L. Michel, *Constraint-based local search* (The MIT Press, Cambridge, 2009)
13. Berthier, D., Constraint Resolution Theories (Lulu.com Publishers, 2011)
14. Ontology-based constraint recognition for free-form service requests. http://www.deg.byu.edu/.../Al-MuhammedEmbley to date
15. C. Gomes, B. Selman, Heavy-tailed phenomena in satisfiability and constraint satisfaction problems. Journ. Autom. Reasoning **24**, 67–100 (2000)
16. R. Barták, M.A. Salido, F. Rossi, New trends in constraint satisfaction, planning, and scheduling: a survey. Knowl. Eng. Rev. 1–24 (2004)
17. Security Ontology. http://securityontology.sba-research.org/ to date
18. A. Herzog, N. Shahmehri, C. Duma, An ontology of information security. Int. J. Inf. Secur. Priv. **1**(4), 1–23 (2007)

19. Ontological Approach toward Cybersecurity in Cloud Computing. http://www.taketaka.com/manuscript/201009sin.pdf
20. C. Blanco et al., A systematic review and comparison of security ontologies, in *Proceedings of International Conference on Availability, Reliability and Security (ARES)*, Barcelona (2008), pp. 813–820
21. A. Souag, Towards a new generation of security requirements definition methodology using ontologies, in *Proceedings of 24th International Conference on Advanced Information Systems Engineering (CAiSE'12) Gdańsk*, Poland, 25–29 June 2012, pp. 1–8
22. C. Torniai et al., Leveraging folksonomies for ontology evolution in e-learning environments, in *Proceedings ICSC'08. Proceedings of the 2008 IEEE International Conference on Semantic Computing* (EEE Computer Society Washington, DC, USA, 2008), pp. 206–213
23. G. Paliouras, D. Constantine, G. Tsatsaronis (eds.), *Knowledge-Driven Multimedia Information Extraction and Ontology Evolution*. Lecture Notes in Computer Science, IX, vol. 6050 (Springer, Berlin, 2011), 245 p
24. Protégé. http://protege.stanford.edu to date

Semiautomatic Telecontrol by Multi-link Manipulators Using Mobile Telecameras

Vladimir Filaretov and Alexey Katsurin

Abstract This chapter describes two methods of semiautomatic position and combined telecontrol by multi-link manipulators using special setting devices (SD), which kinematic schemes are differed from the kinematic schemes of manipulators. For the survey of working space the mobile television cameras are used. Optical axes of these cameras can change its spatial orientations during performance of operations. They can rotate around two mutually perpendicular axes by operator commands. The algorithms of work of computing systems which form the setting signals for drives of all degrees of freedom of manipulators are represented and researched. The results of executed experiments and mathematical simulation of the systems work confirm effectiveness of these methods.

1 Introduction

For rapid and precise execution of some critical operations in the extreme cases the regime of semiautomatic control is frequently used when human-operator sets the motion of gripping device of manipulator or any other operating tool by means of the setting device or control handle and looks only on the image of working space on the screen of telemonitor. Projects in the area of telecontrol by various robots and manipulators are carried out by many researchers in different countries of world [1, 4, 6]. However many key problems are not still solved. In particular for the precise execution of many technological operations in the zone of works it is must be located several television cameras with different orientation of their optical axes or the used television camera must have capability to move in space and to change

V. Filaretov · A. Katsurin (✉)
Robotic Laboratory, Institute of Automation and Control Processes FEB RAS,
Department of Automation and Control, Far Eastern Federal University, Vladivostok, Russia
e-mail: katsurin@mail.ru

V. Filaretov
e-mail: filaret@pma.ru

© Springer International Publishing Switzerland 2016
M. Hadjiski et al. (eds.), *Novel Applications of Intelligent Systems*,
Studies in Computational Intelligence 586, DOI 10.1007/978-3-319-14194-7_11

the orientation of its optical axis. In this case the operator in the process of the setting of grip motions must constantly consider mutual orientation of SD, manipulator and television camera. It leads to the large load on the man. The questions of effective telecontrol by multi-link manipulators using several or one, but mobile telecamera practically have not been examined in the literature.

The approach which makes it possible to consider the current orientation of the optical axis of mobile television camera is proposed in the paper [5]. But the task of the formation of required control signals for manipulator with using SD, when this orientation is changed, is not examined. The papers [2, 3] resolves the problem of telecontrol by multi-link manipulator when the telecamera orientation changes. But the questions of initial co-ordination of the image of manipulator on the monitor screen and the real position of SD are not examined. As a result some motions of this SD can't be executed by manipulator with the concrete kinematical scheme. Some motions are carried out, but with the large errors (in more detail these problems will be described below). Furthermore in the indicated works it is not solved the problem of the elimination of ambiguities which appear during calculation of some trigonometric functions when the inverse task of kinematics is solved.

As a result the problem of developing of new methods and algorithms of semiautomatic telecontrol by multi-link manipulators with a changing of orientation the optical axis of television camera, which is located in the zone of the realization of working operations, is remained the important.

2 Formulation of the Problem

This paper resolves the problem of developing of two methods of semiautomatic position and combined telecontrol by multi-link manipulators using the setting devices, which kinematic schemes are differed from the kinematic schemes of manipulators. But manipulator grip must carry out all motions and take of all positions, given by operator using the SD. In this case it is necessary to receive control laws (setting signals for manipulator drives) and to take into account current orientation of television cameras, which are established in the work zone and formed image on the operator telemonitor in the process of realization of working operations. These cameras can change orientation of their optical axes per rotating about two mutually perpendicular axes. This paper not describes the control signals for orientation of telecameras, but the operator can choose their orientations for reception of the best view of working zone. The solution of the indicated problem must not only considerably to increase productivity and accuracy of operations, performed in the extreme conditions, but also to reduce psychological load on operator.

3 System of Position Telecontrol by the Manipulator

Figure 1 shows the scheme of the semiautomatic telecontrol system by the spatial motion of grip 6 of six-degree manipulator 5 using the six-degree SD 1. In this case the indicated manipulator and SD have different kinematic schemes, which make possible for grip 6 to carry out all motions and to take of all positions, given by operator using the SD. Figure 1 has following designations: $C_i (i = \overline{1,7})$—the systems of coordinates (SC) rigidly connected, correspondingly, with the basis of SD, the basis of SD handle, the SD handle 2, the basis of manipulator, the basis of

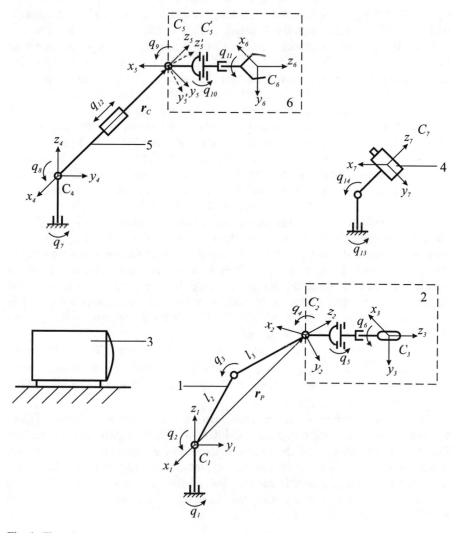

Fig. 1 The scheme of the system of position telecontrol by multi-link manipulator

grip, the grip of manipulator and the body of television camera 4; r_c and r_p—the position vectors, correspondingly, of the basis of manipulator grip in SC C_4 (on Fig. 1 r_c coincides with the telescopic link) and of the basis of SD handle in SC C_1; $q_i(i = \overline{1,14})$—the generalized coordinates of corresponding degrees of freedom of SD, manipulator and television camera; l_2, l_3—the length of corresponding SD links which always lie in one plane.

It should be noted that the longitudinal axes of joints q_8 and q_9, as longitudinal axes of joints $q_i(i = \overline{2,4})$ are always parallel between one another, correspondingly. The axes x_2 and x_5 always coincide with the longitudinal axes of joints q_4 and q_9, correspondingly. The axes z_2 and z_5 always coincide with the link l_3 and the vector r_c, correspondingly, and the axes y_2, y_5 make with the axes x_i and $z_i(i = 2, 5)$ of SC C_2 and C_5 the rights-hand system. The purpose and features of SC C_5' orientation will be explained further with the description of algorithm of system work. The axis z_3 coincides with the longitudinal axis of joint q_6 which is always perpendicular to longitudinal axes of joints q_4 and q_5. The longitudinal axes of joints q_4 and q_5 are always perpendicular between one another. The axes z_6 and y_6 lie in the plane of manipulator grip device. The axis z_6 coincides with the longitudinal axis of joint q_{11} which is always perpendicular to longitudinal axes of joints q_9 and q_{10}. The axes of joints q_9 and q_{10} also are always perpendicular between one another. The axis x_6 makes with two remaining axes of SC C_6 the rights-hand system. The axis x_7 is always parallel to the longitudinal axis of joint q_{14} which is always perpendicular to the vertical longitudinal axis of joint q_{13}. The axis y_7 always coincides with the optical axis of television camera 4 and directs towards its objective. In this case the axis x_7 makes with two remaining axes of SC C_7 the rights-hand system.

For convenience in the subsequent calculations the fixed SC C_1 and C_4 are elevated, correspondingly, to joints q_2 and q_8. Their axes z_1 and z_4 are vertical and they always coincide with longitudinal axes of joints q_1 and q_7, correspondingly. The plane of the screen of monitor 3 is fixed in SC C_1. In this case the motion of SD handle and the motion of the image of manipulator grip on the screen of this monitor are perceived by operator as motions in SC C_1. In accordance with this it is possible to consider that the SC C_1 and C_7 coincide for the operator although they actually coincide only in special cases.

In initial positions of all devices of the system (see Fig. 2) there are $q_i = 0(i = \overline{1, 11}, 13, 14), q_{12\max} = l_2 + l_3$, manipulator and SD are located vertically, and the directions of all axes of SC $C_i(i = \overline{1,7})$ coincide. The axes $z_i(i = \overline{1,7})$ occupy strictly vertical position.

The proposed system of position telecontrol by the manipulator (see Fig. 1) can have two regimes [2, 3]: the regime of stabilization and the regime of tracking. The first regime is auxiliary, and the second is basic. In the regime of stabilization the manipulator is fixed in the space and the operator has the capability to change SD configuration for more comfortable executing of subsequent operations. In the same regime it can occur a changing of orientation of optical axis of telecamera which is established in the work zone.

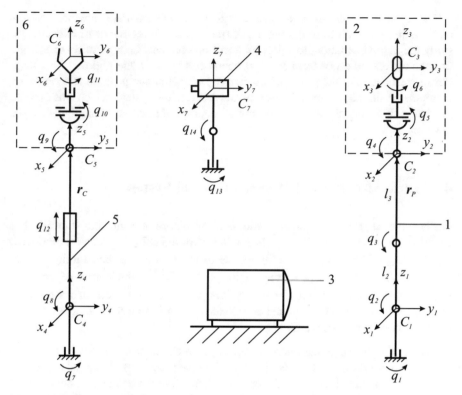

Fig. 2 Initial positions of manipulator, SD and telecamera

According to the difference between the kinematic schemes of manipulator and SD the regime of stabilization makes it possible to execute one additional important function which was absent in the papers [2, 3]. At some mutual orientations of manipulator and SD frequently it was impossible subsequent control of manipulator because of its kinematic limitations or leaving of working zone border. Therefore in the regime of stabilization the operator must not only select convenient for it initial SD configuration but also coordinate it with the current configuration (observed on the screen of telemonitor) and the kinematic possibilities of used manipulator. It is obvious that in this case not only the image of grip but also entire manipulator as a whole must be on the monitor.

At switching moment from the regime of stabilization to the regime of tracking the setting signals for drives of all degrees of freedom of manipulator are formed. These signals provide the co-ordination of its position with the SD orientation and take into account the current orientation of the telecamera optical axis. Subsequently the manipulator repeats any motions determined by SD in the regime of tracking. In this case it is supported the mutual orientation of SD handle and image of manipulator grip on the screen of monitor at any orientation of television camera.

Using this system in nonstandard situations the operator always has the opportunity to switch from the regime of tracking to the regime of stabilization, to study the current situation, to select the most rational new configuration of SD and to continue execution of technological operations again in the regime of tracking. For the realization of the proposed method of semiautomatic position telecontrol by manipulator the algorithm of work of the computing system which ensures the formation of setting signals for drives of all degrees of freedom of manipulator was developed.

4 Work Algorithm of Position Control System

Further the stages of semiautomatic control algorithm will be showed. For describing this algorithm following designations are introduced: $\mathbf{R}_{Xi}, \mathbf{R}_{Yi}, \mathbf{R}_{Zi} \in \mathbf{R}^{3 \times 3}$—correspondingly, the matrices of elementary turnings about axes of coordinates x, y, z to the angle q_i; $\mathbf{A}_j^i \in \mathbf{R}^{3 \times 3}$—the matrix of complex turning consisting of several elementary turnings. The mentioned matrix converts some vector assigned in the rotated coordinate system C_j into the vector assigned in the fixed coordinate system C_i [7].

Stage 1 On this stage the matrix of turning, which connects the fixed coordinate systems C_1 and C_4, is found. As it is mentioned upper the SC C_1 and C_7 coincide visually for the operator. Therefore the connection between SC C_1 and C_4 (or that the same between SC C_7 and C_4) depends only on the current orientation of telecamera (see Fig. 1). As in the regime of tracking the orientation of telecamera does not change so the values of the elements of turning matrix $\mathbf{A}_7^{4*} = \mathbf{A}_1^{4*}$, which will determined on this stage, also does not change. This calculation is carry out once at moment when the system is switching from the regime of stabilization into the regime of tracking. Subsequently elements determined on this stage will be further marked by symbol *.

Using a known rule of transformation of objects coordinates from SCC_7 to C_4 [7] it is simple to show that

$$\mathbf{A}_7^{4*} = \mathbf{R}_{Z13} \cdot \mathbf{R}_{X14} = \begin{bmatrix} c_{13} & -s_{13} & 0 \\ s_{13} & c_{13} & 0 \\ 0 & 0 & 1 \end{bmatrix} \begin{bmatrix} 1 & 0 & 0 \\ 0 & c_{14} & -s_{14} \\ 0 & s_{14} & c_{14} \end{bmatrix}$$
$$= \begin{bmatrix} c_{13} & -s_{13}c_{14} & s_{13}s_{14} \\ s_{13} & c_{13}c_{14} & -c_{13}s_{14} \\ 0 & s_{14} & c_{14} \end{bmatrix}, \tag{1}$$

where c_i, s_i are $\cos(q_i), \sin(q_i)$, correspondingly.

Stage 2   The regime of tracking begins on this stage. In this regime the manipulator must repeat all SD motions. Therefore the stages of the proposed
algorithm beginning from this are carried out cyclically. On this stage
the angles of rotations of SD links (generalized coordinates of SD) are
measured using sensors at current time. The corresponding values of sine
and cosine of these angles are calculated further. The obtained results are
used for the definition elements of the matrices $\mathbf{A}_2^1, \mathbf{A}_3^2$ and the coordinates of the vector $\mathbf{r}_p$.

For obtaining the matrix $\mathbf{A}_2^1$ it is necessary to turn the system of coordinates C_2 first at the angle $(q_2 + q_3)$ about to the turned axis x and then
at the angle q_1 about to the axis z_1 (see Fig. 1). As a result:

$$\mathbf{A}_2^1 = \mathbf{R}_{Z1} \cdot \mathbf{R}_{X2,3} = \begin{bmatrix} c_1 & -s_1 & 0 \\ s_1 & c_1 & 0 \\ 0 & 0 & 1 \end{bmatrix} \begin{bmatrix} 1 & 0 & 0 \\ 0 & c_{2,3} & -s_{2,3} \\ 0 & s_{2,3} & c_{2,3} \end{bmatrix}$$

$$= \begin{bmatrix} c_1 & -s_1 c_{2,3} & s_1 s_{2,3} \\ s_1 & c_1 c_{2,3} & -c_1 s_{2,3} \\ 0 & s_{2,3} & c_{2,3} \end{bmatrix}, \tag{2}$$

where $c_{i,j}, s_{i,j}$ are $\cos(q_i + q_j), \sin(q_i + q_j)$, correspondingly.

For obtaining the matrix $\mathbf{A}_3^2$ the system of coordinates C_3 must be
consecutively turned at the angles q_6, q_5, q_4 about to corresponding axes
z, y, x. The matrix $\mathbf{A}_3^2$ will be presented in the following form:

$$\mathbf{A}_3^2 = \mathbf{R}_{X4} \cdot \mathbf{R}_{Y5} \cdot \mathbf{R}_{Z6} = \begin{bmatrix} c_5 c_6 & -c_5 s_6 & s_5 \\ s_4 s_5 c_6 + c_4 s_6 & -s_4 s_5 s_6 + c_4 c_6 & -s_4 c_5 \\ -c_4 s_5 c_6 + s_4 s_6 & c_4 s_5 s_6 + s_4 c_6 & c_4 c_5 \end{bmatrix}. \tag{3}$$

Counting all generalized coordinates of the SD counterclockwise (in a
positive direction) from its initial position (see Fig. 2), the elements of a
vector $\mathbf{r}_p$ in coordinate system C_1 are determined as:
$x_1 = (l_2 s_2 + l_3 s_{2,3}) s_1, y_1 = -(l_2 s_2 + l_3 s_{2,3}) c_1, z_1 = l_2 c_2 + l_3 c_{2,3}$.

Stage 3   Numerical values of elements of the vector $\mathbf{r}_c = \mathbf{A}_7^{4*} \cdot \mathbf{r}_p$ in the system of
coordinates C_4 are determined on this stage according to known values
of elements of vector $\mathbf{r}_p$. I.e. new position of manipulator in SC C_4,
which defined by operator using the SD, is determined.

Stage 4   The inverse task of kinematics (ITK) for the manipulator is solved. The
numerical values of generalized coordinates of transferred degrees of
freedom of manipulator are determined at current time. They correspond
to current values of elements of the vector $\mathbf{r}_c$. These coordinates determine setting signals for drivers of transferred degrees of freedom of
manipulator.

It is possible to determine scalar of the vector r_c as $|r_c| = |r_p| = q_{12} = \sqrt{x_4^2 + y_4^2 + z_4^2} \geq q_{12\,min}$ (see Fig. 1). Taking into account signs of the corresponding generalized coordinates of the manipulator, all components of vector r_c can be determined using the equations

$$x_4 = q_{12}s_8s_7, \quad y_4 = -q_{12}s_8c_7, \quad z_4 = q_{12}c_8. \tag{4}$$

From last equation of system (4) it is possible to determine coordinate $q_8 = \arccos(z_4/q_{12})$. Further from the first equation of system (4) it would be possible to determine the generalized coordinate $q_7 = \arcsin(x_4/q_{12}s_8)$, but in this case the system degenerates and loses controllability when $q_8 = 0$. For elimination of the indicated situation it is reasonable to determinate the coordinate q_7 by another way.

Earlier already it was stated that in considered system spatial position of the image of manipulator grip 6 on the screen of the telemonitor 3 (see Fig. 1) always coincides with a real position of the SD handle 2 in fixed SC C_1 subject to the current orientation of optical axis of a television camera 4. Thus the orientation of the SD handle in SC C_2 coincide with orientation of manipulator grip in SC C_5' (on Fig. 1 the system of coordinates C_5' is shown by a dotted line, the axes x_5 and x_5' are coincide). SC C_5' is turned relative to SC C_5 at angle $(q_2 + q_3 - q_8)$ about to axis x_5 because of constructive features of the manipulator and the SD. The coincidence of SC C_5' and C_5 will be observed only when $q_3 = 0$. It is possible to determine the transfer matrix $\mathbf{A}_{5'}^4$ from SC C_5' to SC C_4 as:

$$\mathbf{A}_{5'}^4 = \mathbf{R}_{Z7} \cdot \mathbf{R}_{X2,3} = \begin{bmatrix} c_7 & -s_7 & 0 \\ s_7 & c_7 & 0 \\ 0 & 0 & 1 \end{bmatrix} \begin{bmatrix} 1 & 0 & 0 \\ 0 & c_{2,3} & -s_{2,3} \\ 0 & s_{2,3} & c_{2,3} \end{bmatrix}$$

$$= \begin{bmatrix} c_7 & -s_7c_{2,3} & s_7s_{2,3} \\ s_7 & c_7c_{2,3} & -c_7s_{2,3} \\ 0 & s_{2,3} & c_{2,3} \end{bmatrix}. \tag{5}$$

But on the other hand because for operator the systems of coordinates C_1 and C_7 always coincide it is possible to determine values of elements of a matrix $\mathbf{A}_{5'}^4$ per using matrices $\mathbf{A}_2^1$ and $\mathbf{A}_7^{4*}$ calculated at the previous stages of algorithm:

$$\mathbf{A}_{5'}^4 = \mathbf{A}_1^{4*}\mathbf{A}_2^1 = \mathbf{A}_7^{4*}\mathbf{A}_2^1. \tag{6}$$

Thus after determining numerical value of an element a_{11} of matrix $\mathbf{A}_{5'}^4$ using the Eq. (6) and in view of features of calculation of this element using the Eq. (5) it is possible to determine coordinate $q_7 = \arccos(a_{11})$.

Stage 5 The numerical values of generalized coordinates, which assign the orientation of manipulator grip, are determined. At first it is necessary to calculate numerical values of matrix $\mathbf{A}_6^5$ elements.

Because the orientation of the image of manipulator grip 6 on the screen of the telemonitor 3 (see Fig. 1) always coincides with real orientation of the handle of SD 2 and because for operator the systems of coordinates C_1 and C_7 always coincide, the matrix $\mathbf{A}_6^5$ can be presented in form, which is convenient for the decision of a inverse task of kinematics of manipulator:

$$\mathbf{A}_6^5 = \mathbf{A}_4^5 \mathbf{A}_7^{4*} \mathbf{A}_2^1 \mathbf{A}_3^2 = (\mathbf{A}_5^4)^{\mathrm{T}} \mathbf{A}_7^{4*} \mathbf{A}_2^1 \mathbf{A}_3^2, \tag{7}$$

because for orthogonal matrixes of turning, the equality $\mathbf{A}_4^5 = (\mathbf{A}_5^4)^{\mathrm{T}}$ is correct [7]. Matrices $\mathbf{A}_7^{4*}, \mathbf{A}_2^1, \mathbf{A}_3^2$ are determined at stages 1 and 2 of considered algorithm.

The matrix $\mathbf{A}_5^4$ can be determined per taking into account the coordinates q_7, q_8 which were determined on stage 4:

$$\mathbf{A}_5^4 = \mathbf{R}_{Z7} \cdot \mathbf{R}_{X8} = \begin{bmatrix} c_7 & -s_7 & 0 \\ s_7 & c_7 & 0 \\ 0 & 0 & 1 \end{bmatrix} \begin{bmatrix} 1 & 0 & 0 \\ 0 & c_8 & -s_8 \\ 0 & s_8 & c_8 \end{bmatrix}$$
$$= \begin{bmatrix} c_7 & -s_7 c_8 & s_7 s_8 \\ s_7 & c_7 c_8 & -c_7 s_8 \\ 0 & s_8 & c_8 \end{bmatrix}. \tag{8}$$

On the other hand the matrix $\mathbf{A}_6^5$ can be represented as:

$$\mathbf{A}_6^5 = \mathbf{R}_{X9} \cdot \mathbf{R}_{Y10} \cdot \mathbf{R}_{Z11}$$
$$= \begin{bmatrix} c_{10}c_{11} & -c_{10}s_{11} & s_{10} \\ s_9 s_{10} c_{11} + c_9 s_{11} & -s_9 s_{10} s_{11} + c_9 c_{11} & -s_9 c_{10} \\ -c_9 s_{10} c_{11} + s_9 s_{11} & c_9 s_{10} s_{11} + s_9 c_{11} & c_9 c_{10} \end{bmatrix}. \tag{9}$$

In result having calculated all elements of a matrix $\mathbf{A}_6^5$ using the Eq. (7) and taking into account forms of these elements which are received in Eq. (9), it is possible to determine all generalized coordinates of orienting degrees of freedom of manipulator: $q_{10} = \arcsin a_{13}, q_9 = \arccos(a_{33}/c_{10}), q_{11} = \arccos(a_{11}/c_{10})$, where $a_{i,j}$ are numerical values of corresponding elements of the matrix $\mathbf{A}_6^5$. Thus all necessary coordinates of manipulator $(q_7 - q_{12})$ are determined on stages 4 and 5.

Stage 6 If at current time the regime of tracking is assigned, then the transfer to beginning of cycle (to stage 2) is carried out. If the regime of stabilization is assigned, then transfer to this operating mode is carried out.

Table 1 Values of generalized coordinates

Inverse trigonometric functions	Function f	Coordinate q
arcsin	$-\pi \leq f \leq -\pi/2$	$q = -\pi - \arcsin(x)$
	$-\pi/2 \leq f \leq \pi/2$	$q = \arcsin(x)$
	$\pi/2 \leq f \leq \pi$	$q = \pi - \arcsin(x)$
arccos	$-\pi \leq f \leq 0$	$q = -\arccos(x)$
	$0 \leq f \leq \pi$	$q = \arccos(x)$

Table 2 Forms of function f

Generalized coordinate	Form of function f
q_7	$(q_1 + q_{13})$
q_8	$n = l_2 \sin(q_2 + q_{14}) + l_3 \sin(q_2 + q_3 + q_{14})$
q_9	$m = (q_2 + q_3 - q_8) + q_4 + q_{14}$
q_{10}	q_5
q_{11}	q_6

In former algorithm the ambiguity appears when the inverse trigonometric functions are calculated on stages 4 and 5. For its elimination it is necessary to select the calculated angle (generalized coordinate) from that quadrant, in which the current value of the corresponding generalized coordinate is located. For the solution of this problem the special function f was added. This function will determine the quadrant of the corresponding generalized coordinate q. According to properties of arcsine and arccosine the coordinate q can be determined by known value of argument x, taking into account the value of function f (see Table 1). The form of function f for each generalized coordinate of manipulator q_i was determined (see Table 2).

As a result, it is possible to determine values of corresponding functions f at each moment of the time by measured values of generalized coordinates of setting device and telecamera (see Table 2). Further according to the obtained value of function f it is possible to determine the quadrant of the corresponding generalized coordinate and its current value in accordance with Table 1.

5   Simulation of Work of the Position Telecontrol System

For checking of efficiency of the proposed algorithm of semiautomatic position control the simulation of system work was carried out. It was assumed that $l_2 = 0.5$ m; $l_3 = 0.3$ m; the length of vector $\boldsymbol{r}_c$ can vary from 0.2 to 0.8 m; coordinates $q_i(i = \overline{1, 4, 11})$ have ranges of changes $[0, \pi]$; and coordinates q_2, q_3—ranges $[0, \pi/2]$. The laws of variation of generalized coordinates of SD in the time

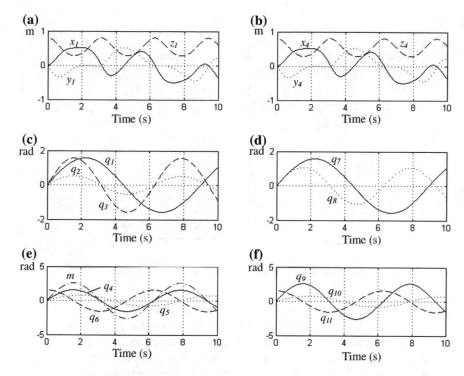

Fig. 3 Results of simulation when $q_{13} = q_{14} = 0 = $ const

have forms: $q_1 = (\pi/2)\sin 0.7t$, $q_2 = (\pi/6)\sin t$, $q_3 = (\pi/2)\sin t$, $q_4 = (\pi/2)\sin t$, $q_5 = (\pi/4)\sin 0.8t$, $q_6 = (\pi/2)\cos t$.

The simulation of the work of system began, when manipulator, SD and television camera was in the initial position (see Fig. 2). The results of simulation when $q_{13} = q_{14} = 0 = $ const are represented on Fig. 3. From Fig. 3a, b it is evident that the coordinates of the vectors $r_p = (x_1, y_1, z_1)^T$ and $r_c = (x_4, y_4, z_4)^T$ always coincide, i.e. the manipulator accurately repeats the spatial motion of SD. In this case (see Fig. 3c, d) the generalized coordinates q_1 and q_7 are equal, and the coordinate q_8 takes the value, which is calculated on the stage 4 of algorithm, and depends on the current values of coordinates q_2 and q_3. Figure 3e, f show that the coordinates q_{10}, q_{11} coincide with coordinates q_5 and q_6, correspondingly, and the coordinate q_9 equal to the angle $m = (q_2 + q_3 - q_8) + q_4 + q_{14}$ which depends on geometric features of manipulator and SD (see Fig. 1).

On Fig. 4 the results of the simulation of system with the turning of the television camera are shown when $q_{13} = \pi/4 = $ const and $q_{14} = 0 = $ const. From Fig. 4a, b it is evident that the turning of television camera is leads to a change of the coordinates of the vector $r_c = (x_4, y_4, z_4)^T$. With this orientation of television camera the coordinate z_4 is not change (it coincides with z_1), but the coordinates x_4, y_4 are change. From Fig. 4d it is evident that the generalized coordinate q_8 is not change, but q_7 is change (see Fig. 3d). Before the regime of tracking start it is necessary to co-ordinate

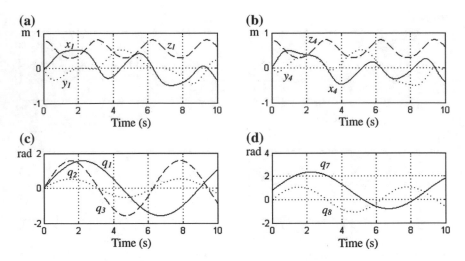

Fig. 4 Results of simulation when $q_{13} = \pi/4 = $ const and $q_{14} = 0 = $ const

the initial configuration of manipulator on the screen of monitor and configuration of SD. In this case the manipulator must be turned at the joint q_7 to the same angle $q_{13} = \pi/4$, because television camera is turned. Further the coordinate q_7 must repeat the law of variation of q_1, which is assigning using the SD, and take into account the initial disagreement, i.e. $q_7 = (q_1 + q_{13})$ (see Fig. 4c, d).

When the proposed approach is used for telecontrol by manipulator using the special SD (see Fig. 1), the some complexities appear if the television camera is turned to the angle q_{14}. If $q_{14} \neq 0$ then the manipulator will not be able to carry out the assigned motions when the coordinate q_1 is change, because it has structural features (see Fig. 1). In this case the turning of SD to the angle q_1 must lead to the turning of manipulator at the joint q_7 to the same angle about to axis, which is turned to the angle q_{14} about to axis x_4. But it is impossible.

For eliminating of indicated negative situation one of the following approaches must be used when television camera is turn to the angle q_{14}:

1. The law of SD motion formulate such as the turning of SD to the angle q_1 is exclude. In this case the manipulator must move in the plane, which is perpendicular to axis of joint q_8.
2. It is provide the special procedure, when the regime of tracking is started with initial disagreement at the angle q_{14} between the configuration of manipulator on the screen of monitor and the configuration of SD. In this case all further motions of SD must be repeated by manipulator accurately.

Thus the results of executed mathematical simulation confirm capacity for work and effectiveness of the proposed method of the synthesis of semiautomatic system of position telecontrol by the manipulator and the work algorithm of computing system.

6 System of Combined Telecontrol by the Manipulator

Figure 5 shows the system of semiautomatic control which is working in a mode of combined control. In this system spatial motion of grip 7 of six-degree manipulator 6 is formed by the operator using the special SD 1 which has three degrees of freedom. The control of grip motion is carried out by its image on the screen of the telemonitor 5. This image is received using the television camera 8, which have two degrees of freedom. Figure 6 has following designations: $C_i (i = \overline{1,6})$—the systems of coordinates rigidly connected correspondingly with the basis of SD, the SD handle 2, the basis of manipulator, the basis of grip, the grip of manipulator and the body of television camera 4; r_c—the position vector of the basis of manipulator grip in SC C_3; $q_i (i = \overline{1,12})$—the generalized coordinates of corresponding degrees of freedom of SD, manipulator and television camera; l_1, l_2—the length of corresponding manipulator links which always lie in one plane.

The fixed systems of coordinates C_1 and C_3 are elevated, correspondingly, to joints q_2 and q_6 as well as in system of position telecontrol. Axes z_1 and z_3 are

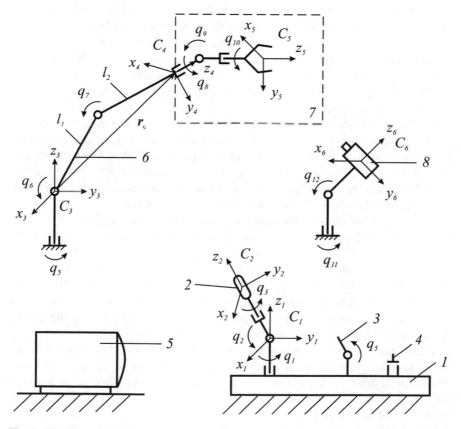

Fig. 5 The scheme of the system of combined telecontrol by multi-link manipulator

Fig. 6 The scheme for
decision ITK of manipulator

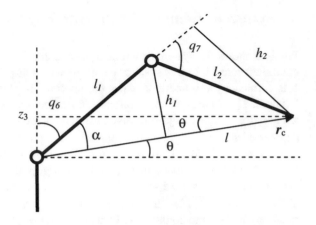

vertical and they always coincide with longitudinal axes of joints q_1 and q_5, correspondingly. The plane of the screen of monitor 5 is fixed in SC C_1. The motion of SD handle and the motion of the image of manipulator grip on the screen of monitor 5 are perceived by operator as motions in SC C_1. In accordance with this it is possible to consider that for operator the SC C_1 and C_6 coincides.

In initial positions of all devices of the system there are $q_i = 0 (i = \overline{1, 12})$, manipulator and SD are located vertically, and the directions of all axes of SC $C_i (i = \overline{1, 6})$ coincide. Axes $z_i (i = \overline{1, 6})$ occupy strictly vertical position.

The proposed system (see Fig. 5) of combined telecontrol by the manipulator has two alternating regimes [2]: regime of grip transfer to the assigned point of space and regime of spatial orientation of grip. In the transfer regime the control of grip motion is carried out by vector of speed. The spatial orientation of grip is provided by position control. In both regimes the control is carried out using the same three-degree SD with a special design (see Fig. 5) simultaneously or separately in time. Switching between regimes and switching on simultaneous performance of these regimes is carried out using the three-position switch 4.

In the regime of transfer the longitudinal axis of the handle 2 always specifies a direction of current motion of grip and coincides with a vector of speed of this motion, which is set by the operator. This axis can have any orientation in space. The value of deviation of setting device 3 (see Fig. 5) from its initial position determines the scalar of this vector of speed. Grip motion in space is carried out by changing of generalized coordinates $q_i (i = \overline{5, 7})$ and the direction of this motion on the screen of telemonitor always coincides with a direction specified by SD handle. At this regime the spatial orientation of grip is kept constant.

· After switching (using the switch 4) at the second regime the automatic co-ordination of orientation of manipulator grip image on the screen of the telemonitor with the current orientation of SD handle 2 is carried out. In other words the coincidence of orientation of SC C_5 axes on the screen of telemonitor with orientation of SC C_2 axes (see Fig. 5) is provided in the second regime at first. Further the operator can change grip orientation by changing spatial orientation of

SD handle 2. When the generalized coordinates $q_i (i = \overline{1,3})$ of SD are change, the generalized coordinates $q_i (i = \overline{8,10})$ of manipulator are change too. But the orientation of grip image on the screen of the telemonitor 5 always coincides completely with the current spatial orientation of SD handle. The vector r_c is kept constant. When the proposed system is working simultaneously at two indicated above regimes, the characteristic point of grip on the screen of the telemonitor is moving in the direction which is specified by the SD handle 2 and with the speed which is set by the operator. The axis z_5 of SC C_5 on the screen coincide with a longitudinal axis of the handle 2, i.e. the manipulator grip always is directed aside the current motion of its characteristic point.

In paper [2] the similar two-regime approach for control by the multi-link manipulator using the similar SD was already offered. But in the second regime (regime of grip orientation) the initial co-ordination of orientation of the grip image on the monitor with the current orientation of the SD handle was not provided. As result the subsequent control of grip orientation using this SD has been considerably complicated.

For the realization of the proposed method of semiautomatic combined tele control by the manipulator the algorithm of work of the computing system has been developed. This algorithm allows forming desirable values of all generalized coordinates of the manipulator (setting signals for drives of all its degrees of freedom) using the three-degree SD (see Fig. 5).

7 Work Algorithm of Combined Control System

Further basic stages of algorithm of combined semiautomatic telecontrol by the six-degree manipulator using the three-degree SD (see Fig. 5) will be showed. At this algorithm the stages 1–4 are realize regime of grip transfer and the stages 6–7—regime of spatial orientation of grip. Consecutive using of all stages 2–8 provides simultaneous performance of both regimes.

Stage 1 On this stage when regime of grip transfer is switching on, the angles of rotations of manipulator links (the generalized coordinates of the manipulator) are measured using the built-in sensors. The corresponding values of sine and cosine of these angles are calculated further. They are used for definition of initial value of coordinates of a vector $r_c = r_{c0}$. All generalized coordinates are counting counterclockwise (in a positive direction) from its initial position. Elements of a vector r_c in SC C_3 are determined using the equations: $x_3 = (l_1 s_6 + l_2 s_{6,7}) s_5$, $y_3 = -(l_1 s_6 + l_2 s_{6,7}) c_5$, $z_3 = l_1 c_6 + l_2 c_{6,7}$. These calculations are carried out once.

Stage 2 Current values of angles q_1 and q_2 of SD handles 2 and deviation angle q_4 of speed setting device 3 are measured. In SC C_1 elements of vector of speed v_1 of manipulator grip motion are determined by formula $v_1 = (q_4 s_1 s_2, -q_4 c_1 s_2, q_4 c_2)^T$.

Stage 3 On this stage the new position of a characteristic point of manipulator
grip is determined. For this the vector of speed v_1 is transferred from SC
C_1 to SC C_3 using the equation $v_3 = A_6^3 \cdot v_1 = A_1^3 \cdot v_1$ at first. Transfer
matrixes $A_6^3 = A_1^3$ connect the fixed SC C_1 and C_3. These matrixes are
equal as for the operator, who is observing the work of manipulator on
the screen of telemonitor, SC C_1 and C_6 always coincide.

Using a known rule of transformation of objects coordinates from SC C_6
to C_3 [7] it is simple to show that

$$A_6^3 = A_1^3 = R_{Z11} \cdot R_{X12} = \begin{bmatrix} c_{11} & -s_{11} & 0 \\ s_{11} & c_{11} & 0 \\ 0 & 0 & 1 \end{bmatrix} \begin{bmatrix} 1 & 0 & 0 \\ 0 & c_{12} & -s_{12} \\ 0 & s_{12} & c_{12} \end{bmatrix}$$

$$= \begin{bmatrix} c_{11} & -s_{11}c_{12} & s_{11}s_{12} \\ s_{11} & c_{11}c_{12} & -c_{11}s_{12} \\ 0 & s_{12} & c_{12} \end{bmatrix}. \tag{10}$$

Further by a vector of speed v_3 the position of grip is determined as
$r_{ci+1} = r_{ci} + v_3 \Delta t$, where $\Delta t = $ const the time interval during which all
calculations corresponding to 2–5 stages of this algorithm are carried
out; r_{ci}—value of a vector r_c on the previous cycle of algorithm performance; r_{ci+1}—new value of a vector r_c.

Stage 4 The inverse task of kinematics of the manipulator is solved. The
numerical values of the generalized coordinates of its three transferred
degrees of freedom $q_i (i = \overline{5,7})$ are determined on the basis of the
received values of elements of a vector r_c. These coordinates are the
setting signals for drives of these degrees of freedom.

It is necessary to note, that a vector r_c with length l and links l_1, l_2 of the
manipulator form a triangle (see Fig. 6). Value l is calculated by equation: $l = \sqrt{x_3^2 + y_3^2 + z_3^2}$, where x_3, y_3, z_3—coordinates of a vector r_c.

The area of this triangle is determined using the equation
$S = 1/2lh_1 = 1/2l_1h_2 = \sqrt{p(p - l_1)(p - l_2)(p - l)}$, where
$p = 0.5(l_1 + l_2 + l)$—half of triangle perimeter, h_1, h_2—its heights.
Values $h_1 = 2\sqrt{p(p - l_1)(p - l_2)(p - l)}/l$ and $h_2 = 2\sqrt{p(p - l_1)(p - l_2)(p - l)}/l_1$ are determined from this equation.

On the other hand from Fig. 6 it is evidently $h_1 = l_1 \sin \alpha, h_2 = l_2 \sin q_7$.
Further it is possible to determine
$q_7 = \arcsin(h_2/l_2) = \arcsin(\frac{2\sqrt{p(p-l_1)(p-l_2)(p-l)}}{l_1 l_2})$.

The coordinate q_6 (see Fig. 6) can be calculated using the equation
$q_6 = \frac{\pi}{2} - (\alpha + \theta) = \frac{\pi}{2} - (\arcsin(\frac{2\sqrt{p(p-l_1)(p-l_2)(p-l)}}{l_1 l}) + \arcsin(z_3/l))$, and
the generalized coordinate q_5 (see Fig. 5)—using the equation:
$q_5 = \arctan(-x_3/y_3)$.

The constructive limitations in degrees of freedom of manipulator exclude an opportunity of occurrence of ambiguity at calculation inverse trigonometric functions when ITK is solved. For this purpose limiting values $q_i(i = \overline{5,7})$ were accepted equal $\pm\pi/2$.

Stage 5 If at current time switch 4 is set at regime of grip transfer, then the transfer to beginning of cycle (to stage 2) is carried out. Otherwise transfer to a stage 6 is carried out.

Stage 6 On this stage the formation of the signals which control orientation of manipulator grip is began. For realization of this regime the transformation from SC C_2 to SC C_1 is carrying out using the matrix $\mathbf{A}_2^1$, i.e. the current orientation of SD handle in SC C_1 is described. For obtaining the matrix $\mathbf{A}_2^1$ the system of coordinates C_2 must be consecutively turned at angles q_3, q_2, q_1 about to corresponding axes z, x, z:

$$\mathbf{A}_2^1 = \mathbf{R}_{Z1} \cdot \mathbf{R}_{X2} \cdot \mathbf{R}_{Z3} = \begin{bmatrix} c_1c_3 - s_1c_2s_3 & -c_1s_3 - s_1c_2c_3 & s_1s_2 \\ s_1c_3 + c_1c_2s_3 & -s_1s_3 + c_1c_2c_3 & -c_1s_2 \\ s_2s_3 & s_2c_3 & c_2 \end{bmatrix}.$$
(11)

Orientation of manipulator grip and connected with it SC C_5 in SC C_4 is described by a matrix $\mathbf{A}_5^4$. Orientation of SC C_4 in SC C_3 is determined by transferred degrees of freedom of manipulator $q_i(i = \overline{5,7})$ and differed from orientation of SC C_2 in SC C_1 (see Fig. 5). However, in this regime the current orientation of the image of manipulator grip (an axis x_5, z_5) on the screen of telemonitor (in SC C_1) must coincide with the current orientation of SD handle (an axis x_2, z_2) in SC C_1. The equality $\mathbf{A}_5^4 = \mathbf{A}_3^4\mathbf{A}_1^3\mathbf{A}_2^1 = (\mathbf{A}_4^3)^{\mathrm{T}}\mathbf{A}_1^3\mathbf{A}_2^1$ (for orthogonal matrixes of turning is correct $\mathbf{A}_3^4 = (\mathbf{A}_4^3)^{\mathrm{T}}$ [7]) should be carried out. The right part of this equality is defined by consecutive transformation from SC C_2 to SC C_1, next—to SC C_3 and next—to SC C_4.

The matrix $\mathbf{A}_1^3$ is determined on stage 3 of algorithm, and the matrix $\mathbf{A}_4^3$ can be determined by values of the generalized coordinates $q_i(i = \overline{5,7})$ which are already calculated on stage 4:

$$\mathbf{A}_4^3 = \mathbf{R}_{Z5} \cdot \mathbf{R}_{X6,7} = \begin{bmatrix} c_5 & -s_5 & 0 \\ s_5 & c_5 & 0 \\ 0 & 0 & 1 \end{bmatrix} \begin{bmatrix} 1 & 0 & 0 \\ 0 & c_{6,7} & -s_{6,7} \\ 0 & s_{6,7} & c_{6,7} \end{bmatrix}$$
$$= \begin{bmatrix} c_5 & -s_5c_{6,7} & s_5s_{6,7} \\ s_5 & c_5c_{6,7} & -c_5s_{6,7} \\ 0 & s_{6,7} & c_{6,7} \end{bmatrix}.$$
(12)

Stage 7 On this stage the values of three last generalized coordinates which determine spatial orientation of manipulator grip are calculated. It is obvious that the matrix $\mathbf{A}_5^4$ can be determined as:

$$\mathbf{A}_5^4 = \mathbf{R}_{Z8} \cdot \mathbf{R}_{X9} \cdot \mathbf{R}_{Z10}$$

$$= \begin{bmatrix} c_8 c_{10} - s_8 c_9 s_{10} & -c_8 s_{10} - s_8 c_9 c_{10} & s_8 s_9 \\ s_8 c_{10} + c_8 c_9 s_{10} & -s_8 s_{10} + c_8 c_9 c_{10} & -c_8 s_9 \\ s_9 s_{10} & s_9 c_{10} & c_9 \end{bmatrix}. \quad (13)$$

From this equation for elements of matrix $\mathbf{A}_5^4$ it is possible to calculate the generalized coordinates of all orienting degrees of freedom of manipulator: $q_9 = \arccos(a_{33})$, $q_8 = \arcsin(a_{13}/s_9)$, $q_{10} = \arcsin(a_{31}/s_9)$, where a_{ij} are numerical values of corresponding elements of matrix $\mathbf{A}_5^4$, which are calculated on stage 6. As it visible from equation for q_8, q_{10}, the system loses controllability when $q_9 = 0$. For elimination of this situation it is necessary to maintain a condition $q_9 \neq 0$ during control process.

In former algorithm the ambiguity appears when inverse trigonometric functions are calculated. For its elimination it is necessary to limit values of generalized coordinates. Coordinates q_8, q_{10} must have ranges of change $[-\pi/2, \pi/2]$; coordinate q_9—a range $(0, \pi]$.

The generalized coordinates, which are calculated on this stage, are the setting signals for drives of corresponding degrees of freedom of manipulator grip. As result the image of this grip on the screen of telemonitor completely coincides with orientation of SD handle in SC C_1.

Stage 8 If at current time only the regime of spatial orientation of grip is assigned, then the transfer to stage 6 is carried out. If only the regime of grip transfer is assigned, then the transfer to stage 1 is carried out. Otherwise transfer to stage 2 is carried out.

8 Simulation of Work of the Combined Telecontrol System

For checking efficiency of the proposed algorithm of semiautomatic combined control the simulation of system work was carried out. It was assumed that $l_1 = 0.5\,\text{m}; l_2 = 0.3\,\text{m}$; coordinates $q_i(i = \overline{1,3})$ have ranges of change $[-\pi, \pi]$; coordinates $q_i(i = \overline{5,8}\,,\,\overline{10,12})$—ranges $[-\pi/2, \pi/2]$; coordinate q_4—a range $[0, \pi/2]$; and coordinate q_9—a range $(0, \pi]$. Dynamics of the manipulator was not taken into account during simulation. The size Δt at the third stage of the proposed algorithm is equal $\Delta t = 0.01$ s.

During the simulation the process of the work of system is divided into two stages: in the time interval 0–5 s the system works in the regime of transfer, and in the interval 5–10 s in the regime of grip orientation. At initial moment the generalized coordinates of manipulator are accepted equal: $q_5 = q_7 = 0, q_6 = \pi/2$, and SD handle 2 and speed setting device 3 (see Fig. 5) are turned to the angles: $q_1 = -\pi/4, q_2 = -\pi/3, q_3 = 0, q_4 = \pi/9$. For this regime the laws of variation of generalized coordinates of SD in the time have forms: $q_1 = (-\pi/5)\sin 0.5t$, $q_2 = (\pi/3)\sin 0.3t$, $q_3 = (-\pi/6)\cos 0.5t$.

The results of simulation when $q_{11} = q_{12} = 0$ are represented on Figs. 7, 8 and 9. The laws of variation of generalized coordinates of SD are shown on Fig. 7a. Figure 7b shows the changes of generalized coordinates of the transferred degrees of freedom of manipulator $q_i(i = \overline{5,7})$ and length l of vector r_e during the transfer regime. But manipulator motion is ceases when it reaches the boundary of working zone ($l_{max} = 0.8$ m) during the motion in the assigned direction. In the orientation regime ($t = 5$–10 s) vector r_e does not change when the position of SD handle does change. The laws of variation of generalized coordinates of the orienting degrees of freedom of grip q_8, q_9, q_{10} are shown on Fig. 7c. The grip orientation in the regime of transfer ($t = 0$–5 s) does not change (see Fig. 7), but the laws of variation of generalized coordinates of grip in the orientation regime are differed from the laws

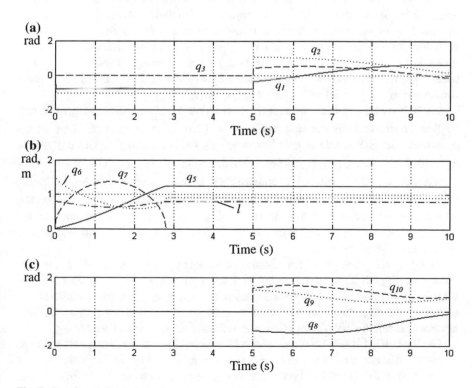

Fig. 7 Results of simulation when $q_{11} = q_{12} = 0 = $ const

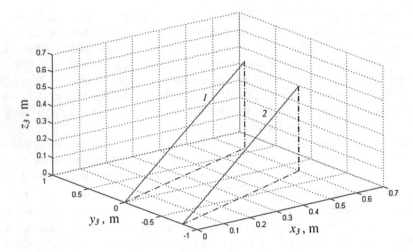

Fig. 8 Motion of characteristic point of grip when $q_{11} = q_{12} = 0$

of SD handle motion. Because the orientation of the basis of manipulator grip (SC C_4) is not coinciding with the orientation of the basis of SD handle (SC C_1).

On Fig. 8 line 1 is shown. This line coincides with vector of speed $\mathbf{v_1}$, which is assigned by SD handle in SC C_1. The results of simulation showed that the vector of speed $\mathbf{v_3}$ of the motion of the grip characteristic point in SC C_3 completely coincides with vector of speed $\mathbf{v_1}$ when $q_{11} = q_{12} = 0 = \text{const}$ (television camera is located in the initial position). On same figure the trajectory of motion of characteristic point of manipulator grip indicated by number 2. This characteristic point moves strictly in parallel to vector of speed $\mathbf{v_1}$ ($\mathbf{v_3}$).

For a visual assessment of mutual orientation of SD handle and grip during motion of manipulator the unit vector $e_2 = (0, 0, 1)^\text{T}$ was entered. This vector coincides with SD handle orientation in SC C_2 (with an axis z_2). The unit vector $e_5 = (0, 0, 1)^\text{T}$ in SC C_5 coincides with direction of axis z_5. The vector e_2 is transferred to SC C_1 using the equation $e_{21} = A_2^1 e_2$, and the vector e_5 to SC C_3—$e_{53} = A_4^3 A_5^4 e_5$, and to SC C_1—$e_{51} = A_3^1 A_4^3 A_5^4 e_5$, where e_{21} is the vector, which shows orientation of SD handle in SC C_1, e_{51} and e_{53}—the vectors, which show orientation of manipulator grip in SC C_1 (on the screen of telemonitor) and in SC C_3, correspondingly.

The graphs of changes of coordinates of vectors e_{21}, e_{51} and e_{53} in the time are shown on Fig. 9. Television camera is fixed ($q_{11} = q_{12} = 0$) and SD handle is moved. In this case, when the coordinates of the vector e_{21} (triangular markers) are changing, the coordinates of vectors of orientation of manipulator grip e_{53} and e_{51} always coincide between each other and with the coordinates of vector e_{21}.

On Figs. 10, 11 and 12 the results of the simulation of system work with the turned telecamera are shown, when $q_{11} = q_{12} = \pi/4 = \text{const}$. In this case the laws of variation of SD generalized coordinates in time are not changed (see Figs. 7a and

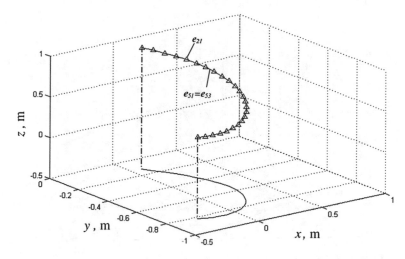

Fig. 9 Coordinates of the vectors which are showing the orientation of the SD and grip when $q_{11} = q_{12} = 0$

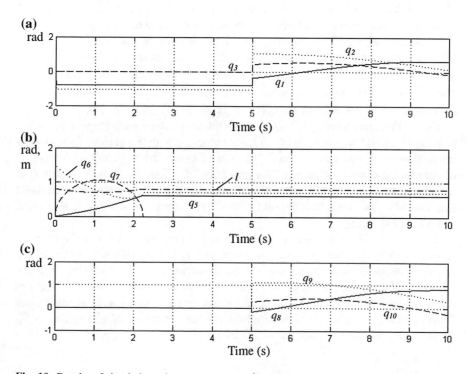

Fig. 10 Results of simulation when $q_{11} = q_{12} = \pi/4 = \text{const}$

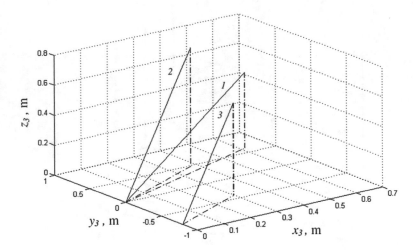

Fig. 11 Motion of characteristic point of grip when $q_{11} = q_{12} = \pi/4$

10a). But the laws of variation of generalized coordinates $q_i(i = \overline{5,7})$ are changed, because television camera had new orientation. In this case the manipulator reaches the boundary of working zone ($l_{\max} = 0.8$ m) at another time—2.2 s (2.8 s on Fig. 7b). The laws of variation of generalized coordinates of manipulator grip $q_i(i = \overline{8,10})$ in the orientation regime (see Fig. 10c), when the orientation of television camera changed, are differed from the laws of motion of SD handle and from the laws of variation of coordinates $q_i(i = \overline{8,10})$ (see Fig. 7c).

Figure 11 shows lines *1, 2, 3*. Line *1* coincides with the vector of speed $\boldsymbol{v_1}$, which is assigned by SD handle in SC C_1. Line *2* coincides with the vector of speed $\boldsymbol{v_3}$ of the motion of manipulator grip in SC C_3. Line *3* shows the trajectory of the motion of characteristic point of manipulator grip in SC C_3. The direction of vector $\boldsymbol{v_3}$ is differing from the vector $\boldsymbol{v_1}$ when the orientation of television camera is changed (see Fig. 11). The characteristic point of manipulator grip moves strictly in parallel to vector of speed $\boldsymbol{v_3}$.

The graphs of changes of coordinates of vectors $\boldsymbol{e}_{21}, \boldsymbol{e}_{51}$ and $\boldsymbol{e}_{53}$ in the time are shown on Fig. 12, when $q_{11} = q_{12} = \pi/4 = $ const. The coordinates of vectors $\boldsymbol{e}_{53}$ and $\boldsymbol{e}_{51}$ does not coincide, when the coordinates of vector $\boldsymbol{e}_{21}$ are changing. Because the orientations of SC C_3 and C_1 does not coincide, when television camera is turned. In this case the coordinates of vectors $\boldsymbol{e}_{21}$ and $\boldsymbol{e}_{51}$ are coincide as before, and the orientation of the image of manipulator grip on the screen of telemonitor in SC C_1 coincides with the orientation of SD handle.

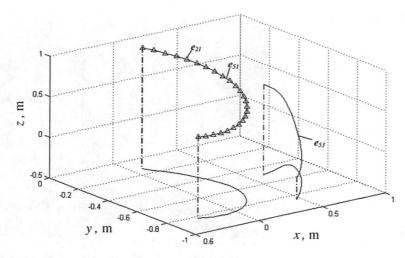

Fig. 12 Coordinates of the vectors which are showing the orientation of the SD and grip when $q_{11} = q_{12} = \pi/4$

9 Experimental Research of System of Telecontrol by the Manipulator

The equipment PHANTOM Omni and Premium 1.5 (SensAble Technologies, Inc.) have been used as setting devices and manipulator. It was used two setting devices: first for control of the manipulator, second for change of orientation of a television camera (Fig. 13). The location of the mobile camera and two identical manipulators (during the experiments one of them has been switched off) is shown on Fig. 14. The mobile camera is established on the robot miniCRANE 5.1 (Robots and Design, Co. Ltd.), which provide turn of the camera around two perpendicular axes.

Real robotic equipment which used for experimental researches has kinematic schemes differing from schemes in Figs. 1 and 5. Figure 15 shows the scheme of this telecontrol system by the spatial motion of manipulator 4 using the special setting device 1. Figure 15 has following designations: $C_i (i = \overline{1,5})$—the systems of coordinates (SC) rigidly connected, correspondingly, with the basis of SD, the basis of SD handle, the basis of manipulator, the characteristic point of manipulator (the endpoint of last link of manipulator) and the body of television camera 3; r_m and r_{sd}—the position vectors of the characteristic point of manipulator in SC C_3 and of the basis of SD handle in SC C_1; $q_i (i = \overline{1,8})$—the generalized coordinates of corresponding degrees of freedom of SD, manipulator and television camera; l_1, l_2—the length of SD links; l_3, l_4—the length of manipulator links; 2—monitor.

Feature of this system is that the SD and the manipulator are located inversely from each other (in initial position of telecamera on Fig. 15 $q_7 = \pi$). The axis x_5

Fig. 13 Operator place

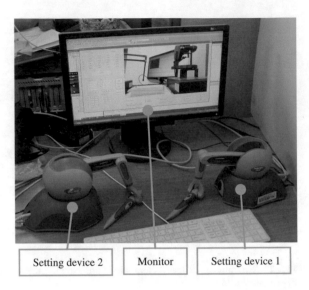

Setting device 2   Monitor   Setting device 1

always coincides with the optical axis of camera 3. The axis z_5 makes with other axes of SC C_5 the rights-hand system. For convenience in the subsequent calculations the fixed SC C_1 and C_3 are elevated, correspondingly, to joints q_2 and q_5. The plane of the screen of monitor 2 is fixed in SC C_1. In this case the motion of SD handle and the motion of the image of manipulator on the screen of this monitor are perceived by operator as motions in SC C_1. In accordance with this it is possible to consider that the systems of coordinates C_1 and C_5 coincide for the operator although they actually coincide only in special cases.

The experimental system of position telecontrol by the manipulator (see Figs. 13, 14 and 15) also has two regimes: the regime of stabilization and the regime of tracking. In the regime of stabilization the manipulator is fixed in the space and the operator has the capability to change SD configuration and orientation of optical axis of telecamera. In the regime of tracking the manipulator repeats any motions formed by SD with taking into account the current orientation of the telecamera optical axis.

For the realization of experimental system of semiautomatic telecontrol by manipulator the algorithm of work of the computing system which ensures the formation of setting signals for drives of all degrees of freedom of manipulator was developed similarly indicated above algorithms.

Fig. 14 Working space

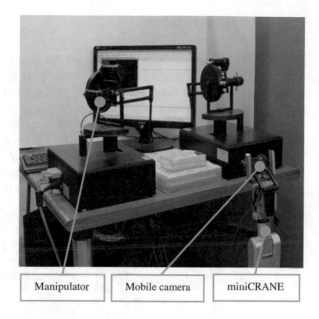

| Manipulator | Mobile camera | miniCRANE |

Fig. 15 The scheme of the experimental system of telecontrol by manipulator

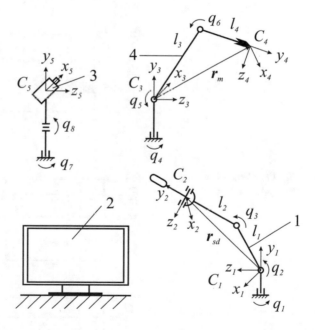

10   Experiment Results

During the experiments the operator moved the handle of the setting device from
any initial position in parallel a monitor plane. Thus the characteristic point of the
manipulator on the monitor screen moved in the same direction (perpendicularly
optical axis of a television camera) at any orientation of a television camera.

On Figs. 16, 17 and 18 results of experiments are shown at the turned television
camera ($q_7 = 5\pi/6 = $ const, $q_8 = 0 = $ const). As initial positions and mutual orientation of the SD and the manipulator do not coincide, during the movement the
generalised co-ordinates of the SD (q_1, q_2, q_3) and the manipulator (q_4, q_5, q_6) and
consequently vectors r_m and r_{sd} also do not coincide (Figs. 16 and 17).

Fig. 16 Generalized
coordinates of the
manipulator and SD

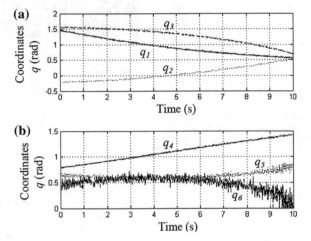

Fig. 17 Elements of a
vectors r_{sd} and r_m

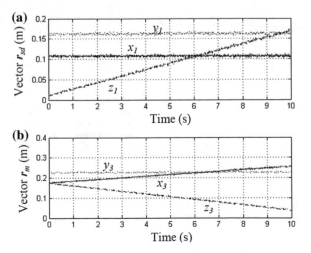

Fig. 18 Motion of SD and characteristic point of manipulator

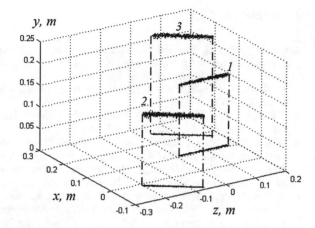

On Fig. 18 spatial trajectories of movement of the SD and a characteristic point of the manipulator are shown at the turned television camera. On Fig. 18 the line 1 coincides with a trajectory of motion of SD handle in SC C_1. The line 2 coincides with direction of SD motion in SC C_3, i.e. line 2 assigns the direction of manipulator motion in SC C_3. The line 3 shows a trajectory of motion of characteristic point of manipulator in SC C_3. Consequently the manipulator on the screen of the operator monitor moved strictly in the direction, which is set using the SD at any orientation of the camera.

11 Conclusions

The proposed methods of semiautomatic position and combined control by manipulators make it possible to automatically consider the spatial orientation of television cameras during the execution of technological operations. In this case the psychological load on human-operator is reduced because he doesn't need to constantly consider in his action the current orientation of telecameras. Also operator always has a possibility to select the convenient orientation of television cameras and position of SD handles, which kinematic schemes are differed from the kinematic schemes of manipulators. Results of the executed experiments and mathematical simulation confirm effectiveness of the proposed systems of telecontrol by multi-link manipulators and the work algorithms of computing systems. Engineering realization of the proposed methods of position and combined telecontrol doesn't cause principle difficulties because in this case it is necessary to execute some additional mathematical calculations only.

Acknowledgements This work was financed by the Russian Foundation for Basic Research grants.

References

1. C. Changhwan, S. Yongchil, J. Seungho, K. Seungho, Tele-operated fuel handling machine manipulation robot for the nuclear power plants, in *Proceedings of the 6th Asian Control Conference*, Bali, Indonesia (2006), pp. 983–987
2. V. Filaretov, Y. Alekseev, A. Lebedev, *Control Systems of Underwater Robots* (Krugly god, Moscow, 2001)
3. V.F. Filaretov, G.P. Kihney, F.D. Yurchik, About one method of telecontrol by manipulator, Izvestiya vuzov. Elektromahanika **3**, 94–98 (1992)
4. D. Lee, M.W. Spong, Passive bilateral control of teleoperators under constant time-delay, in *Proceedings of the 16 IFAC World Congress*, Prague, Czech Rep (2005)
5. T. Murao, H. Kawai, M. Fujita, Passivity-based dynamic visual feedback control with a movable camera, in *Proceedings of 16 IFAC World Congress*, Prague, Czech Rep (2005)
6. R. Tarca, I. Pasc, N. Tarca, F. Popentiu-Vladicescu, Remote robot control via internet using augmented reality, in *Proceedings of the 18-th DAAAM International Symposium "Intelligent Manufacturing & Automation"*, Zadar, Croatia, (2007), pp. 739–740
7. S.L. Zenkevich, A.S. Yuschenko, *The Bases of Control by Manipulate Robots* (MGTU N.E, Bauman, 2004)

Vision-Based Hybrid Map-Building and Robot Localization in Unstructured and Moderately Dynamic Environments

Sherine Rady

Abstract This work focuses on developing efficient environment representation and localization for mobile robots. A solution-approach is proposed for hybrid map-building and localization, which suits operating environments with unstructuredness and moderate dynamics. The solution-approach is vision-based and includes two phases. In the first phase, the map-building reduces the domain of extracted point features from local places through an information-theoretic analysis. The analysis simultaneously selects the most distinctive features only. The selected features are further compressed into codewords. The uncompressed features are also tagged with their metric position. In such a way, a unified map is created with hybrid data representation. In the second phase, the map is used to localize the robot. For fast topological localization, features extracted from the local place are compared to the codewords. To extend the localization into a metric pose, triangulation is executed hierarchically for the identified topological place with the use of the positional metric data of features. To ensure accurate position estimate, the dynamics of the environment are detected through the spatial layout of features and are isolated at the metric localization level. The proposed map-building and localization solution enables for a fast hybrid localization without degenerating the accuracy of localization.

1 Introduction

Map-building and localization are fundamental topics in mobile robots research. The solution approaches are classified into topological and metric in accordance with the type of map employed. Topological maps are well-suited for autonomous navigation especially indoors, while metric maps conform to specific tasks that need

---

S. Rady (✉)
Information Systems Department, Faculty of Computer and Information Sciences,
Ain Shams University, Cairo 11566, Abbassia, Egypt
e-mail: srady@cis.asu.edu.eg; sherine.rady@ziti.uni-heidelberg.de

© Springer International Publishing Switzerland 2016
M. Hadjiski et al. (eds.), *Novel Applications of Intelligent Systems*,
Studies in Computational Intelligence 586, DOI 10.1007/978-3-319-14194-7_12

precision. Most of the current metric map-building and localization approaches are probabilistic. They apply Simultaneous Localization and Mapping (SLAM) while using Extended Kalman Filter (EKF) for relative positioning [8] or Monte Carlo Localization (MCL) for global localization [6, 17]. Probabilistic approaches basically rely on the motion model of the robot. Triangulation and trilateration of pre-located landmarks are another category of solutions for the robot global localization problem [5, 16, 23].

Despite the abundance of current map-building and localization approaches, some difficulties arise with their application in unstructured and dynamic environments.

From the map-building perspective, the problem of which data to employ in the map is still a challenging open-question. Structured features can be easily identified in some operating environments, but they might be lacked in other densely cluttered and unstructured environments. Local point feature extraction [2, 10] have proven to be suitable for modeling unstructured complex scenes, as well as objects. The problem of those techniques is the large number of local points, and consequently features, produced. Some applications have employed these high-dimensional descriptors, directly the way they are extracted by their algorithms, for the purposes of robot localization. Examples are those in [1, 7, 13, 19, 20].

From the localization perspective, the current difficulties lie in the performance of the metric approaches when applied on a large scale. The performance suffers from both high uncertainties and high computational complexity. The uncertainties exist due to the perceptual aliasing, data correspondence and possible environment dynamics, while the high complexity is obviously an escalation with the increased size of environment. Recently, hierarchical localization frameworks have been introduced in which the topological and metric paradigms are fused together. The frameworks offer hybrid solutions with lower complexity for large-scale metric localization, because the searchable space is confined to a prior identified topological location. However, the efficiency of topological localization in those frameworks is critical, since the robot actions based on a misrecognized topological location can be fatal. Practically, the hierarchical localization has been applied differently. For e.g. Voronoi graphs have been constructed with localizing Bayesian filters in [9, 18] and appearance vision-based recognition is fused with probabilistic filters for the update of filter's sample weights in [11, 22]. Moreover, different abstractions of features established another form of hierarchical processing such as in [3, 12, 20]. These implementations start with a fast initial feature matching (e.g. global descriptors), followed by a second matching for the filtered result using another feature set (e.g. local descriptors) to reach a single solution at the end.

In this paper, a hybrid map-building and localization solution-approach is proposed. The map-building employs vision-based local feature extraction with additional attention to the combined quality and size of extracted features. For this purpose, an information-theoretic analysis is involved, for evaluating the information content of features. The analytical procedure generates a map consisting of a reduced feature set that minimizes the aliasing and correspondence problems, and hence achieves both localization speed and accuracy. Additionally, the paper proposes a hierarchical top-down localization using a unified hybrid map. The reduced features obtained from the information-theoretic analysis are stored in two different resolutions. The first resolution is coarse and is obtained by compressing the features using a dictionary compression component. This 'feature form' comprises a non-geometric descriptor and is used for fast topological localization. The second resolution is higher, in which the features are non-compressed and possess an additional corresponding geometric descriptor. More specifically, the features are tagged with their corresponding position information. Based on an identified topological solution (which can be executed on a higher localization abstract level), the geometric data are used in a lower extended localization level, where a triangulation technique is used to obtain a metric position estimate of the robot. For ensuring the accuracy of the metric position estimate, a criterion for the dynamics detection of environment is proposed and is applied before the triangulation execution.

The rest of the paper is organized as follows: First, the hybrid map-building and localization approach, together with a structured solution are introduced in Sect. 2. The modules and structural components of the map-building and localization are described in Sects. 3 and 4 respectively. Section 5 shows the practical experimentation and the approach evaluation. Finally, Sect. 6 concludes the paper.

2 Map-Building and Localization Solution-Structure

Figure 1 introduces the structure of the proposed vision-based map-building and localization solution. The map-building concept is given in Fig. 1a, while the detailed solution structure in Fig. 1b.

The map-building concept processes the environment space and features on both the topological and metric levels. It employs a local feature extraction (see Fig. 1a) to suit densely-cluttered environments or those lacking clear structure. Besides the common feature extraction process, the map-building relies on two additional processes: feature evaluation and feature compression. The feature evaluation will evaluate the features' contribution to the reduction of uncertainties of topological localization, and hence filter the most distinguishing ones for the location. The feature compression will compress the high-dimensional pool of the evaluated filtered set. Finally, the compressed data which relate descriptors to the topological location, together with their tagged metric position data will form a unified hybrid map. This map is seen as possessing two different data resolutions, course and fine, which can resolve the robot position at two localizing scales; topological and metric.

It is worth saying that most of the current map-building implementations use feature extraction as a straight-forward process. However, for the suggested implementation, the newly introduced processes of evaluation and compression will contribute to generating an information-rich and compact map. Such map will provide high localization accuracy with fewer features, which will consequently induce computational savings in the space memory and processing time of the localization.

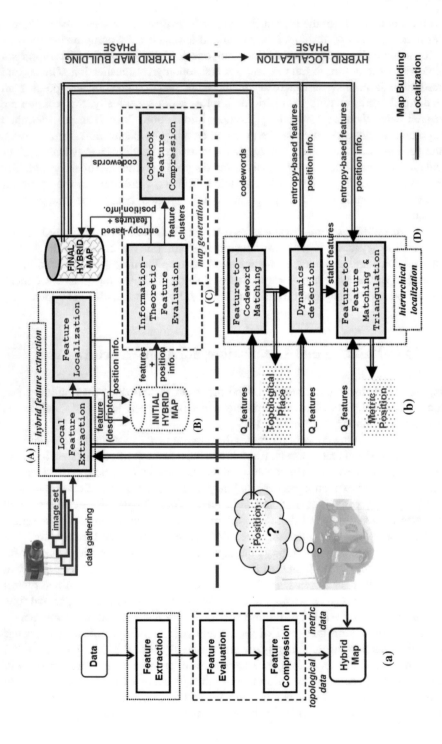

Fig. 1 **a** Map-building concept. **b** Hybrid map-building and localization solution-structure

The proposed solution structure in Fig. 1b outlines two phases: a 'hybrid map-building' phase and a 'hybrid robot localization' phase. The map-building phase uses the concept in Fig. 1a by combining topological and metric data and constructing the hybrid map using the same reduced feature set. Two modules are employed here; a hybrid feature extraction module and a map generation module. They are outlined on the figure as modules (A) and (C) respectively. Module (A) consists of a local feature extraction component and a feature localization component. The components represent and deliver two distinct feature representations in order to resolve the robot location at the two localizing scales; a coarse topological place and a precise metric position. The delivered outputs of the module are a feature descriptor[1] obtained by the local feature extraction component and a corresponding feature's position information obtained by the feature localization component. These data are stored in the temporary storage (B) for further offline processing to obtain an optimized reduced feature-map, which is executed by module (C). Module (C) employs components for the feature evaluation and feature compression. The first filters the most relevant set from the extracted features by applying an information-theoretic method. The second reduces the size of the filtered features further through a Codebook compression technique. Eventually, the generated data by both components (entropy-based features, position information and codewords) form the final hybrid map which is installed on the robot platform for the localization.

The robot localization phase includes a hybrid localization module (D), which uses the final hybrid map data in addition to extracted features from the current robot view to determine the robot's position (see Fig. 1b). The localizing module affords localizing the robot to a topological place, which is done through a feature-to-codeword matching component using the codewords. The module also extends the localization hierarchically to a precise metric value, if desired, through the feature-to-feature matching and triangulation component that uses the entropy-based features and their position information. Prior to the metric localization estimation, the possible environment dynamics and topological-level mismatches are detected and eliminated by a dynamics detection component. This is to ensure the accuracy of metric solution.

The previously introduced modules will be explained in more details in the two subsequent sections: Hybrid map-building and hybrid localization.

3 Hybrid Map-Building

This phase contains three modules, namely, (A) Hybrid feature extraction module; (B) Initial map, and (C) Map generation module.

---

[1]The feature descriptor will be outlined as feature in the paper for easiness.

3.1 Hybrid Feature Extraction Module

Local point feature extraction techniques, such as the Scale Invariant Feature Transform (SIFT) [10] or the Speeded-Up Robust Features (SURF) [2] are suggested for the local feature extraction component. The techniques are distinguished by richness of their descriptors and robustness for detection under different scale and view point, beside other transformations and distortions. Most importantly, they can map environments with poor structure. Their disadvantage is the high computational power involved in the extraction and matching. The map generation module described in the third subsection contributes to solving such a problem by speeding up the matching. In our implementation, the SIFT is used for its popularity.

The second feature localization component calculates the position of the extracted features in a global frame of reference by applying triangulation. Different positions of the moving robot are used as input for the triangulation execution, together with the corresponding measured bearings by the camera (i.e. angles between the unknown camera heading direction and the features which are identified in camera in pixels). The mathematical equations for the triangulation are explained in the localization section (Sect. 4).

In our work, it was desired to quantify the robot localization metric errors precisely, with less measurement error obtained in the map. Therefore, robot positions were provided from an accurate external positioning system (Krypton K600). Those robot positions were fed to the features' localization component in the triangulation. We note that the same system has been used for referencing (i.e. generating the ground truth data) as well.

3.2 Initial Map

In map-building phase, the robot moves in the operating environment to collect image data. The environment is divided into a set of discrete topological places (nodes) related to each other by neighborhood interconnectivities. Such nodes can be assigned automatically by generating them at equidistant distances traveled by the robot. A second option is to assign the new node at the region exhibiting a variation in the scene details which can be monitored by the dynamic change of the tracked features in the camera view. A third alternative is to instruct the robot to assign the new node, which can be done by learning or manually by pressing a button at the needed place. The last option has been employed in our current implementation.

After the robot collects data at each node, an initial hybrid map is constructed. The map consists of nodes with local features that possess both non-geometric and geometric data obtained by components of module (A). The non-geometric data part is represented by the Gradient descriptor of SIFT and is used as the characterization of the topological nodes. The geometric data part is represented by the tagged position of the features in the nodes and is used together with the non-geometric part as the data resolving the robot's metric location. The initial hybrid map does not specifically contain the most informative data. Excessive or redundant data may exist, which increase the localization uncertainty and computational overhead. Therefore, the map undergoes further processing by module (C), the map generation module, in order to extract the possible minimum data that maximize the localization accuracy and enhance its computational efficiency.

3.3 Map Generation Module

For a more efficient-use map, the initial map is processed by the analytical approach introduced in [14, 15]. The approach makes use of an information-theoretic analysis for feature evaluation and a Codebook for feature compression. An entropy-based evaluating criterion is defined and combined with a clustering technique to support additional compression for the evaluated data. Accuracy combined with computational efficiency has been achieved for robot indoor topological localization [14] and under severe illumination conditions [15].

The evaluation process aims at minimizing the uncertainty associated with the topological localization, and in other terms maximizing the topological node recognition. This is performed by the evaluating criterion to filter out the most discriminative data at each node for the initial map constructed in the previous section. The filtered set is named entropy-based features as identified in Fig. 1b. The evaluation is implemented in a way to supply cluster information to a second Codebook component (feature clusters). The Codebook compresses the evaluated feature clusters further resulting in a format called the codewords (see Fig. 1b). Hence, codewords are considered the compressed version of the entropy-based features and contain non-geometric data part only.

In the frame of the information-theoretic analytical approach, every topological node is initially described by a set of local features, and moreover the same feature set is compressed to define a set of local codeword candidates. Compression is done by clustering the features extracted in every node using k-means clustering algorithm. Next, the whole feature pool of the nodes is sampled to determine the true variation (distribution) of the features in the high-dimensional space. Afterwards, a conditional-entropy-based criterion is applied to evaluate the contribution of features to uniquely identify a local codeword candidate. This correspondingly measures the features' contribution to identify a specific node because the codewords are defined locally. The criterion has the following form [4]:

$$H(O|f_i) = - \sum_k P(o_k|f_i) \cdot \log_2 P(o_k|f_i) \qquad (1)$$

where o_k represents a local codeword candidate and f_i represents a true variation sample; $k = 1,\ldots, \Omega$; Ω is the total number of codeword candidates, and $i = 1,\ldots, \Psi$; Ψ

is the total number of feature samples which are termed feature categories. Finally, the filtering process is conducted, where the codeword candidates and their corresponding uncompressed features exhibiting relatively large entropy values are discarded, while those exhibiting low entropy values are selected for the final map. Eventually, the final hybrid map contains those reduced feature set, which are called 'entropy-based features' and 'codewords', and which are descriptions of 'Gradient and positional data' and additional 'compressed Gradient data' respectively.

4 Hybrid Localization

The hybrid localization module includes three components which are processed sequentially: a topological matching component, a dynamics detection component and finally a feature identity resolving and triangulation component. The first component matches features extracted from the current camera view with the map to identify the current topological node. Extracted features are matched against the codewords using the Cosine distance and the majority votes to indicate the best matched topological place(s).

The second dynamics detection component detects the environment dynamic features. Dynamic features exist when some environment objects are relocated to new places instead of their previously registered ones. The component also manages to detect possible mismatched feature pairs that occur at the topological matching component. Failing to specify the identity of a feature because of multiple occurrence or similarity metrics is the cause of such mismatch (i.e. data association or correspondence problem). In another view, the identity of a dynamic object can be correctly identified; however features of such object carry false information for robot localization, and hence are not desired. One can see that those features should be eliminated from the map if it is required to maintain static features only for metric localization purpose. From another perspective, those features can be helpful for topological place recognition and therefore can still be stored. They should be only excluded at the metric level.

Both mismatched and dynamic features can be detected using the spatial relationships between features. A distance measure is proposed, in order to classify the undesired features as outliers, which has the form:

$$d_m = \left(\sqrt{(X_m - X'_m)^2 + (Y_m - Y'_m)^2} \right) \tag{2}$$

where (X_m, Y_m) and (X'_m, Y'_m) are the location of the mth matched pair in the current image view and the image of the recognized topological node respectively; $m = 1$,

2, ..., M; and M is the number of matched features between the two images. With a simple splitting method like clustering or histogram generation for the calculated distances D = $\{d_1, d_2, ..., d_m\}$, outliers are easily encapsulated, and hence excluded. A small size for the number of clusters or the histogram bin size (e.g. 3–4) is sensitive enough to detect the outliers.

The third component of the localization module triangulates the identified static features to estimate the metric robot position by using the non-compressed map data. The component processes the localization hierarchically after the topological node is identified. The identity of the node features is first resolved by matching the extracted features to the entropy-based features. Next, the metric position data of features are triangulated. Two triangulation methods are tested. They both require the observed bearing angles measured by the camera. The first method is the Geometric Triangulation which calculates geometric and trigonometric relation-ships to deduce the unknown position in 2-dimensional space [5]. The second method is the iterative Gauss-Newton method which applies a photogrammetric projective model in the 3-dimensional space to calculate the position estimate. The projective model relates the 3-D Cartesian coordinates of features in the real world to their corresponding 2-D coordinates as projected on the image plane. The fundamental equations describing such model, and which are used in the majority of applications, are called the Collinearity equations. They relate the object point and its corresponding image point to the perspective camera center point. Projections on the image plane in the x and y directions are proportional to the horizontal and vertical bearings respectively (x_a, y_a), and can be described by the so-called Collinearity equations [21]:

$$x_a = -f \frac{r_{11}(X_A - X_C) + r_{21}(Y_A - Y_C) + r_{31}(Z_A - Z_C)}{r_{13}(X_A - X_C) + r_{23}(Y_A - Y_C) + r_{33}(Z_A - Z_C)} \tag{1}$$

$$y_a = -f \frac{r_{12}(X_A - X_C) + r_{22}(Y_A - Y_C) + r_{32}(Z_A - Z_C)}{r_{13}(X_A - X_C) + r_{23}(Y_A - Y_C) + r_{33}(Z_A - Z_C)} \tag{2}$$

where (X_A, Y_A, Z_A) is the feature position data, (X_C, Y_C, Z_C) is camera position located by its perspective center, f is the camera focal length, and r_{ij} are elements of the rotation matrix describing the transformation between the camera and the global world coordinate systems.

$$\substack{W \\ C}R = R_x(\omega) \cdot R_y(\phi) \cdot R_z(\kappa) = \begin{bmatrix} r_{11} & r_{12} & r_{13} \\ r_{21} & r_{22} & r_{23} \\ r_{31} & r_{32} & r_{33} \end{bmatrix} \tag{4}$$

$$= \begin{bmatrix} \cos\phi\cos\kappa & -\cos\phi\sin\kappa & \sin\phi \\ \cos\omega\sin\kappa + \sin\omega\sin\phi\cos\kappa & \cos\omega\cos\kappa - \sin\omega\sin\phi\sin\kappa & -\sin\omega\cos\phi \\ \sin\omega\sin\kappa - \cos\omega\sin\phi\cos\kappa & \sin\omega\cos\kappa + \cos\omega\sin\phi\sin\kappa & \cos\omega\cos\phi \end{bmatrix} \tag{5}$$

The estimation of camera position is described as a non-linear least squares problem. A residual cost function is defined to minimize the sum of the squared error between the actual measured positions of the features in the image and their calculated values from the model Eqs. (1) and (2). The function has the form:

$$r = \sum_{i=1}^{n} \left(\hat{x}_a^i + f\frac{U^i}{W^i} \right)^2 + \left(\hat{y}_a^i + f\frac{V^i}{W^i} \right)^2 \qquad (6)$$

where $(\hat{x}_a, \hat{y}_a)$ are the measured feature position in the camera frame in pixels, n is the number of features, and:

$$U^i = r_{11}(X_A^i - X_C) + r_{21}(Y_A^i - Y_C) + r_{31}(Z_A^i - Z_C) \qquad (7a)$$

$$V^i = r_{12}(X_A^i - X_C) + r_{22}(Y_A^i - Y_C) + r_{32}(Z_A^i - Z_C) \qquad (7b)$$

$$W^i = r_{13}(X_A^i - X_C) + r_{23}(Y_A^i - Y_C) + r_{33}(Z_A^i - Z_C) \qquad (7c)$$

Solving Eq. (6), the camera position is calculated, from which the robot position can also be determined. Since the camera is fixed on the robot platform, a fixed offset vector $\mathbf{q_{off}}$ and a rotation matrix $\mathbf{R}$ defines the relationship between the robot body frame and camera frame.

$$^W\mathbf{q_r} = {^W\mathbf{q_c}} - \mathbf{R} \cdot \mathbf{q_{off}} \qquad (8)$$

where $^W\mathbf{q_r}$ and $^W\mathbf{q_c}$ define the position vectors of the robot and the camera in the global world coordinate frame respectively.

For the localization solution-approach, few remarks are worth to mention. First, the localization computational complexity is lower because of the hierarchical processing. Second, the localization is robust against uncertainties in the node recognition and data correspondences. This is because: (1) Features used in the map building are selected based on properties of features relevance and node distinguishness measured by the evaluation criterion (i.e. ambiguities minimized). (2) The feature-to-feature matching in the metric localization is confined to a single topological node, and hence, the probability of mismatches and correspondence errors is reduced in comparison to the case of matching with all features in the global map.

A final remark concerns the feature localization component in Fig. 1b, which localizes the features in the map building phase. The same proposed triangulation solution in this section is used for the feature localization component to estimate the location of features in the 3-dimensional space. The solution to Eqs. (1) and (6) is easier this time, since the Collinearity equations are not function of the non-linear parameters of $^W_C R$, which are known this time. The feature's location (X_A, Y_A, Z_A) are directly calculated given the camera position and orientation vectors besides the bearing measurements.

5   Experimentation

The experimenting robot is a Pioneer P3-DX platform. It is equipped with a ring of 8 forward ultrasonic sensors engaged in a single behavior for collision avoidance. A structure has been built up on the robot to support a webcam and a laptop at suitable height and view. The webcam is Logitech 4000 pro of resolution 640×480. It is fixed on a rotating mechanism where panoramic images are constructed through stitching to provide rich characterization for the scenes of the places. The mechanism allows 2–4 images to be captured and stitched using the extracted local features. The attached laptop has an Intel core Duo 2.4 GHz processor, with 3 GB RAM, windows operating system and Matlab development environment. Only the vision-based system is engaged in the map-building and localization behavior under test.

The testing is conducted for a 120 m^2 area in the indoor office environment of the Automation department at Heidelberg University. The area constitutes 7 places accounting for the map topological nodes. It spans 5 office sections in 3 rooms (2 double-office rooms and one single-office room), a meeting hall with a joint kitchen and an exit/stairs view. For data gathering and initial map building, the robot wanders at the different places in several rounds performed at apart timings. During these timings, the office environment places are changed through the moved, missed or newly added objects. The gathered data are used as the training dataset for the information-theoretic evaluation and Codebook modules, and their filtered outputs form the final hybrid map which is installed on the robot.

In a similar manner, a second dataset is collected from the environment in several rounds for the testing. To localize the robot in the testing phase, the robot had to visit places, executes sequential rotation actions to build a panoramic view for each place. Next, the extracted features from the panoramic view under query are compared to the map. Using the Cosine distance, apart distanced feature matches are discarded and the matched topological place is determined through the majority votes acquired. This is followed by the dynamics detection and hierarchical metric localization using either the Geometric Triangulation or Iterative Gauss-Newton methods.

Firstly, we show results of employing the entropy-based feature set generated from feature evaluation as explained in Sect. 3.3. We experimented with different values for the parameter Ψ (number of feature categories) in Eq. (1), between 10 and 1000 for more than 35,000 features in the dataset, and monitored the effect of eliminating features of relatively high entropy values. Figure 2 shows the localization accuracy versus different percentages of elimination of high-entropy features in the training dataset, for $\Psi = 10, 100, 800$ and 3936. The plot indicates that high cluster variation, indicated by Ψ, shows less accuracy ($\Psi = 3936$). This identifies that the data undergo extra division and lose meaningful information content. For less Ψ values, the plots almost maintain constant performance before it starts decreasing at 64 % elimination.

Fig. 2 Average localization
accuracy versus features
elimination

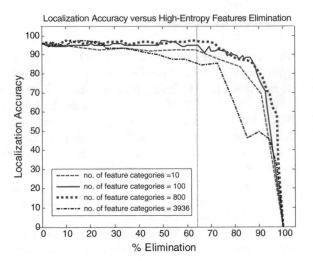

Localization Accuracy versus High-Entropy Features Elimination

The plot still maintains this high accuracy performance up till 72 % for $\Psi = 800$, indicating the best obtained performance at this parameter value. Hence, the results conclude that eliminating up till 64 % of the features as high-entropy features has insignificant effect on the localization accuracy. This means 64 % of the features do not share efficiently in the candidate codeword or place classification, but are rather an overhead on the localization.

Secondly, results of employing the Codebook for the topological localization are demonstrated on two different Codebook examples: the first is obtained from a low-entropy feature set reduced to 74 % of the original features, and has 3845 entries. The second is based on a feature set reduced to 36 %, and has 2739 entries.

Figure 3 shows the performance of the Codebook examples for the training dataset compared to the performance of both the original SIFT and the entropy-based features. The localization performance is shown as a retrieval performance measured by the Precision-Recall relationship, in which the Precision stands for the localization accuracy. For the two used Codebooks, the graph shows similar Precision values to the original SIFT features obtained at lower Recall values. For higher Recall values, the Precision is higher using the Codebook, which implies that successive retrievals for the place is obtained more efficiently than the original SIFT features which result in more false matches. The reason for that is because of the feature evaluation.

Figure 4 shows the two Codebooks performance for the testing dataset. Matching using the Codebook also induces a similar localization performance to the original SIFT features. As shown in the figure, the incorporation of the evaluation component has also influenced the top matches of the Codebooks compared to the original SIFT features. Precision recorded 100 % up till 35 % Recall value versus 20 % Recall value for original SIFT features.

The Codebook does not only maintain a similar performance to the original features, but also has a significant reduction effect on the size of the topological

Fig. 3 Average retrieval and localization performance for the training dataset using Codebook based on **a** 26 % feature reduction (3845 Entries). **b** 64 % feature reduction (2739 Entries)

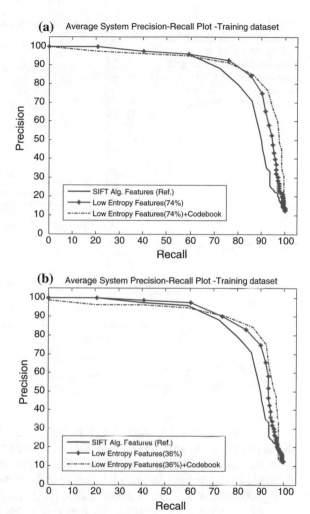

(a)   Average System Precision-Recall Plot -Training dataset

(b)   Average System Precision-Recall Plot -Training dataset

features stored in the map. This reduction is about 90 % compared to original SIFT features, and about 80 % compared to the entropy-based features (SIFT-map $\sim$35K total features; 74 % low-entropy features map $\sim$26K total features; 36 % low-entropy features map $\sim$13K total features). Using the Codebook has also resulted in a reduction for the matching time by almost 90 %.

Table 1 summarizes the topological localization performance for the initial SIFT map, the entropy-based map and the Codebook (based on 36 % low-entropy feature set) for different performance indices with the savings attained. The topological localization for these results has been set to be the Precision value at 60 % Recall. This allows to judge the localization not only by its first best match retrieval but also through a distribution for the retrievals.

Fig. 4 Average retrieval and
localization performance for
testing dataset using
Codebook based on **a** 26 %
feature reduction (3845
Entries). **b** 64 % feature
reduction (2739 Entries)

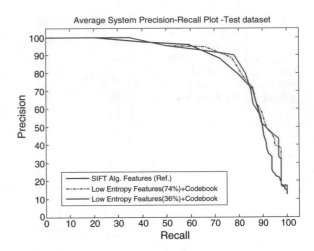

Average System Precision-Recall Plot -Test dataset

— SIFT Alg. Features (Ref.)
---- Low Entropy Features(74%)+Codebook
— Low Entropy Features(36%)+Codebook

Table 1 Topological localization performance: performance index versus method

Performance index	Method		
	SIFT	Entropy-based map	Code-book
Number of features/image	1000	370	67
Localization accuracy (%)	95.5	96.9	93.6
Memory space reduction (%)	–	68	93.3
Localization time reduction (%)	–	54.6	78.3

The previous results show the topological localization performance. It is
extended here with the dynamics detection and metric triangulation components, in
order to evaluate the metric localization. Recall that the hierarchical metric localization is executed by a first place matching at the topological level, and next by a
triangulation execution at the second metric level using the detected features in the
topological solution provided by the first level. The filtered data based on pre-
serving 36 % low-entropy feature set is used. In this set, the topological data size is
2.67 MB with an average number of features per place equal to 67. The metric data
size is 12.81 MB, with an average number of 320 features per place. Both constitute
the data to be processed in the hierarchical localization.

Figure 5 shows a test example for the dynamics detection component. The two
images represent the current camera view as seen by the robot and the best matched
place image retrieved by the topological matching component. The detected paired
feature matches are shown on the images. Using a histogram of three bins for
Eq. (2), three outliers are detected, from which two are mismatched features, while
the third is a feature for a relocated object. This test example has been recognized
with a topological localization accuracy of 100 % at 60 % Recall.

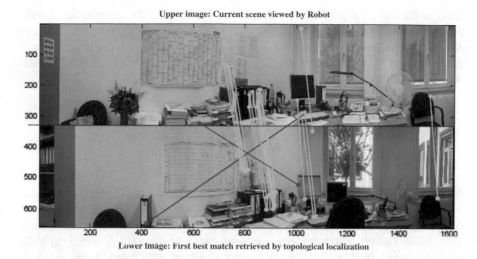

Upper image: Current scene viewed by Robot

Lower Image: First best match retrieved by topological localization

Fig. 5 Dynamics detection component. From matching, features are classified into inliers (shown in *cyan*) and outliers (shown in *magenta*). The figure shows two identified dynamic features and a single mismatch. The *upper image* is the current camera view by the robot, and the *lower image* is the identified topological node

Table 2 Metric localization performance: performance index versus method

Performance index	Method	
	Geometric triangulation method	Iterative Gauss-Newton method
Average positional x-error (cm)	10.0167	6.1363
Average positional y-error (cm)	34.4958	4.0432
Average rms error (cm)	35.9207	7.3486
Average orientation error (°)	5.0393	2.0789
Average execution time (ms)	22.008	69.842
Maximum execution time (ms)	23.974	149.36

Table 2 summarizes the metric errors of the triangulation component for the Geometric Triangulation and the Iterative Gauss-Newton methods, in which the Krypton K600 system has been used to generate the Ground Truth data. The Geometric Triangulation applies only three-feature triangulation, while the Iterative Gauss-Newton triangulation uses more than three features. That is why the localization errors in the latter are less than the former. The iterative Gauss-Newton method executes longer than the Geometric Triangulation because it includes iterations and requires initialization. The initialization is set to the origin of the local coordinate frame of each topological node, with zero orientation value. Though not used in our case, but the first identified metric location can be used to initialize the following position estimate in time as the robot moves on its trajectory. This will speed up the localization obviously. The obtained metric performance results indicate an average root mean square positional error in the range of 7–36 cm and average orientation errors of about 2–5°.

5.1 Localization Performance Under Acquisition Disturbances

The topological localization has been additionally tested under different acquisition disturbances. Figure 6 shows image examples with disturbed acquisition conditions. The images have been acquired in two situations. In the first situation, the robot is driven with relatively high velocity, the matter that introduced high vibrations in the camera. In the second situation, the robot is driven with lower velocity which yielded less camera vibrations.

These situations, accordingly, influenced the quality of image stitching. It is necessary to report the effect of those disturbances and the consecutive stitching quality on the localization performance. Two webcams with two different resolutions have been additionally tested: Logitech 4000 pro (320 × 240) and Logitech C600 (640 × 480). This is because the image resolution is expected to have a consecutive effect on the image stitching quality.

The introduced high vibrations lead to bad-quality stitching as clearly illustrated by Fig. 6a, c. Figure 6a, b show the panoramic view constructed by the low resolution camera from 3-image stitching procedure, while Fig. 6c and d show the view constructed by the high resolution camera from 6-image stitching procedure. The images show that higher resolution images undergo low-quality stitching when the camera exhibits severe vibrations.

The localization performance under the disturbed acquisition conditions is summarized in Table 3. Employing a low-resolution sensor provides more robust localization towards the low-quality stitching caused by severe vibrations (Table 3a). The Codebook performance is slightly affected by the stitching quality than the entropy-based features which shows more robustness. On the other hand, employing a high-resolution sensor provides less robust localization towards the low-quality stitching and severe vibrations (Table 3b). However, employing the high-resolution sensor shows good performance when vibrations are less and stitching is of high-quality as emphasized in bold in Table 3. This high-quality stitching can be obtained through a stabilized platform. Stabilization is guaranteed by acquiring the image shots when the robot executes a rotation control then pauses, instead of acquiring the shots while the robot platform is in motion.

Table 4 shows the time needed for stitching images together (2, 3, 4 and 6). Stitching with the low-resolution camera required almost 1 s for every single stitching. This time is four times doubled with the high-resolution camera since the

(a)

(b)

(c)

(d)

Fig. 6 Examples for disturbed acquisition conditions. Stitching quality is influenced by sensor vibration and resolution: **a** 3-image stitching (320 × 240), high vibrations. **b** 3-image stitching (320 × 240), low vibrations. **c** 6-image stitching (640 × 480), high vibrations. **b** 6-image stitching (640 × 480), low vibrations

feature extraction generates larger number of features. Experimentations have demonstrated stitching up till 3 images are enough for capturing the information of the scene.

Table 3 Localization performance with disturbed acquisition conditions

Parameter/method	Less vibrations/good-quality stitching		Severe vibrations/low-quality stitching	
	Entropy-based map	Codebook	Entropy-based map	Codebook
(a) Logitech 4000 pro (320 × 240)				
Avg. matching time (s)	2.8475	0.8532	2.1094	0.5726
Avg. localization time (s)	4.009	2.9060	4.1622	2.6254
Best match localization (%)	100	50	**100**	**100**
Distribution localization[a] (%)	100	50	**100**	**50**
(b) Logitech C600 (640 × 480)				
Avg. matching time (s)	9.8620	2.5056	8.6820	1.9469
Avg. localization time (s)	17.9329	10.5709	16.6734	10.0156
Best match localization (%)	**100**	**100**	25	25
Distribution localization[a] (%)	**100**	**75**	75	25

[a]60 % Recall localization

Table 4 Stitching performance

	Stitching time(s)			
	2 images	3 images	4 images	6 images
Logitech 4000 pro (320 × 240)	1.0955	2.0528	2.9486	4.3467
Logitech C600 (640 × 480)	4.3019	8.0709	11.5317	16.3182

Stitching time versus resolution and number of images

6 Conclusions

This paper presents a vision-based solution-approach for map-building and robot localization. The solution-approach is suitable for unstructured and moderately dynamic environments. The approach is capable of localizing a mobile robot on both topological and metric levels. A hybrid map construction is proposed, based on maintaining distinctive features to provide recognition accuracy with less computational overhead. The map stores two data formats, coarse and fine, in order to resolve the topological location and the metric position separately. The proposed hybrid map building has demonstrated to provide substantial savings in the map size. The map, used along within a proposed hierarchical processing for the metric localization, has proven to provide combined accuracy and computational efficiency for localization.

References

1. H. Andreasson, T. Duckett, Topological localization for mobile robots using omnidirectional vision and local features, in *IFAC Symposium on Intelligent Autonomous Vehicles*, Lisbon, Portugal (2004)
2. H. Bay, A. Ess, T. Tuytelaars, L.V. Gool, SURF: speeded up robust features. Comput. Vis. Image Underst. **110**(3), 346–359 (2008)
3. J. Courbon, Y. Mezouar, L. Eck, P. Martinet, Efficient hierarchical localization method in an omnidirectional images memory, in *IEEE International Conference on Robotics and Automation (ICRA)* (2008), pp. 13–18
4. T.M. Cover, J.A. Thomas, Elements of Information Theory. Wiley Series in Telecommunications (1991)
5. J.S. Esteves, A. Carvalho, C. Couto, Position and orientation errors in mobile robot absolute self-localization using an improved version of the generalized geometric triangulation algorithm, in *IEEE International Conference on Industrial Technology (ICIT)* (2006), pp. 830–835
6. D. Filliat, J.-A. Meyer, Map-based navigation in mobile robots: I. A review of localization strategies. J. Cogn. Syst. Res. **4**, 283–317 (2003)
7. L. Goncalves, Di E. Bernardo, D. Benson, M. Svedman, J. Ostrowski, N. Karlsson, P. Pirjanian, A visual frontend for simultaneous localization and mapping, in *IEEE International Conference on Robotics and Automation (ICRA)* (2005), pp. 44–49
8. J.J. Leonard, H.F. Durrant-Whyte, Mobile robot localization by tracking geometric beacons. IEEE Trans. Robot. Autom. **7**(3), 376–382 (1991)
9. B. Lisien, D. Morales, D. Silver, G. Kantor, I. Rekleitis, H. Choset, Hierarchical simultaneous localization and mapping, in *IEEE/RSJ International Conference on Intelligent Robots and Systems (IROS)* (2003), pp. 448–453
10. D.G. Lowe, Distinctive image features from scale invariant keypoints. Int. J. Comput. Vision **2**(60), 91–110 (2004)
11. E. Menegatti, M. Zoccarato, E. Pagello, H. Ishiguro, Image-based Monte-Carlo localisation without a map, in *Proceedings of 8th Conference of the Italian Association for Artificial Intelligence* (2003a), pp. 423–435
12. F. Menegatti, M. Zoccarato, E. Pagello, H. Ishiguro, Hierarchical image-based localisation for mobile robots with Monte-Carlo localisation. *European Conference on Mobile Robots* (2003b), pp. 13–20
13. A.C. Murillo, J.J. Guerrero, C. Sagüés, Surf features for efficient robot localization with omnidirectional images, in *IEEE International Conference on Robotics and Automation (ICRA)* (2007), pp. 3901–3907
14. S. Rady, E. Badreddin, Information-theoretic environment modeling for efficient topological localization, in *Proceedings of 10th International Conference on Intelligent Systems Design and Applications (ISDA)* (2010), pp. 1042–1046
15. S. Rady, A. Wagner, E. Badreddin, Building efficient topological maps for mobile robot localization: An evaluation study on COLD benchmarking database, in *Proceedings of IEEE/RSJ International Conference on Intelligent Robots and Systems (IROS)* (2010), pp. 542–547
16. S. Rady, A. Wagner, E. Badreddin, Hierarchical localization using entropy-based feature maps and triangulation techniques, in *Proceedings of IEEE International Conference on System, Man and Cybernetics (SMC)*, Istanbul, Turkey (2010), pp. 519–525
17. S. Thrun, W. Burgard, D. Fox, *Probabilistic Robotics* (MIT Press, Cambridge, MA, 2005)
18. S. Tully, H. Moon, D. Morales, G. Kantor, H. Choset, Hybrid localization using the hierarchical atlas. in *IEEE/RSJ International Conference on Intelligent Robots and Systems (IROS)* (2007), pp. 2857–2864
19. C. Valgren, A. Lilienthal, Sift, surf and seasons: long-term outdoor localization using local features, in *European Conference on Mobile Robots (ECMR)*, Freiburg, Germany (2007), pp. 1–6

20. J. Wang, H. Zha, R. Cipolla, Coarse-to-fine vision-based localization by indexing scale-Invariant features, in *IEEE Transactions of Systems, Man, and Cybernetics* (*SMC*) (2006), pp. 413–422
21. S.S. Welch, R.C. Montgomery, M.F. Barsky, The spacecraft control laboratory experiment optical attitude measurement system. Technical memorandum 102624, NASA (1991)
22. J. Wolf, W. Burgard, H. Burkhardt, Robust vision-based localization by combining an image retrieval system with Monte Carlo localization. IEEE Trans. Rob. **21**(2), 208–216 (2005)
23. D.K.C. Yuen, B.A. MacDonald, Vision-based localization algorithm based on landmark matching, triangulation, reconstruction, and comparison. IEEE Trans. Rob. **21**(2), 217–226 (2005)

Innovative Fuzzy-Neural Model Predictive Control Synthesis for Pusher Reheating Furnace

Goran S. Stojanovski, Mile J. Stankovski, Imre J. Rudas
and Juanwei Jing

Abstract This chapter is largely based on the paper "Pusher Reheating Furnace Control via Fuzzy-Neural Model Predictive Control Synthesis" presented at IEEE IS 2012 in Sofia, Bulgaria. A design of innovative fuzzy model-based predictive control for industrial furnaces has been derived and applied to the model of three-zone 25 MW RZS pusher furnace at Skopje Steelworks. The fuzzy-neural variant of Sugeno fuzzy model, as an adaptive neuro-fuzzy implementation, is employed as a predictor in a predictive controller. In order to build the predictive controller the adaptation of the fuzzy model using dynamic process information is carried out. Optimization procedure employing a simplified gradient technique is used to calculate predictions of the future control actions.

Keywords Fuzzy neural networks · Fuzzy model predictive control · Optimization · Set-point control · Time-delay processes

1 Introduction

The comprehension philosophy of real-world dynamical processes as a non-separable inter-play of the three fundamental natural quantities of energy, matter, and information, where energy and matter are the carriers of information governing the former ones suites best the thermal systems [1, 2]. In thermal systems

G.S. Stojanovski (✉) · M.J. Stankovski
Faculty of Electrical Engineering and Information Technologies, Rugjer Boshkovikj 18, Skopje 1000, Macedonia
e-mail: goranst@feit.ukim.edu.mk

I.J. Rudas
Institute of Intelligent Engineering Systems, Óbuda University, Budapest H-1034, Hungary

J. Jing
College of Information Science and Engineering Institute of Control Theory and Navigation Technology, Northeastern University, Liaoning 100006, China

© Springer International Publishing Switzerland 2016
M. Hadjiski et al. (eds.), *Novel Applications of Intelligent Systems*,
Studies in Computational Intelligence 586, DOI 10.1007/978-3-319-14194-7_13

such as high-power, multi-zone furnaces the complexity of energy conversion and transfer processes in seems to be ideally suited to the quest for new designs of improved control and supervision strategies. Typical steel-mill slab-pusher furnaces as in Skopje Steelworks [1–3] have high installed power and are operated with gas/oil fuel in a heavy-duty field environment, and also they require considerable maintenance support.

The operational pace is imposed by production scheduling of the slab rolling plant subsequent to the furnace. Main disturbances come along with the line-speed and flow-rate of heated mass as well as the hot gas and air flows inside the reheat processing space; varying slab/ingot dimensions and grades in lower-grade steel processing also act as disturbances; the disturbances are non-stationary stochastic. The economic operation of such a furnace is rather delicate. The overall control task is to drive the slab heating to the desired thermodynamic equilibrium and regulate it as well as to maintain the required temperature profile throughout the furnace. The economic reasons, in particular the energy cost (fuel consumption) per product unit, and operational restrictions make necessary the plant operation optimization.

For this purpose, industrial practice of steelworks demonstrated the alternative optimizations via combined advanced controls: (i) an optimally designed controller to perform both the regulatory and supervisory tasks (e.g., see [4]); and (ii) an optimization expert system is combined with executive regulatory controls to perform both tasks (e.g., see [5]). The first alternative leads to designs that are difficult to implement and maintain tuned, whereas the second one requires a costly development of an appropriate heuristic expert system. It is therefore that a simple alternative strategy employing model-based predictive set-point optimization along with the local standard PI controls via the generalized predictive control theory [6] for cost-effective operation and practical maintenance service of steel-mill furnaces was derived [2].

In this paper, another alternative control system design with two-layer architecture employing T-S fuzzy predictors [7], which seems rather competitive to the optimization expert [5] based ones. The here proposed Fuzzy Model Predictive Control (FMPC) design is a type of model-predictive control strategy in which the prediction process model is a fuzzy or fuzzy-neural model. The control system architecture is constructed of two main sub-systems: a predictor in terms of a dynamic fuzzy model of the thermal process, which calculates the predictions of the process outputs; and a controlling optimizer, which optimizes the process by minimizing the cost function and generating appropriate process controls. In turn, the overall system control is a hybrid one. Further in Sect. 2, a presentation of the newly derived technique is elaborated. Section 3 presents and discusses the simulation results. Concluding remarks along with a relevant discussion and references are given thereafter.

2 Control System Architecture and Fuzzy MPC Strategy

Pusher furnace thermodynamics are processes that have very low-frequency dynamics. In addition, they are highly non-linear and long time-delay processes. It is therefore that sophisticated and computationally expensive control strategies can be applied, including the Model-based Predictive Control (MPC) strategies (Fig. 1; consulting [8], is prerequisite for the work reported here). In designing predictive controllers most complex part is the identification of the corresponding suitable process model to serve for predictions. In due time different types of MPC applications have been developed employing different types of models for output predictions. These include traditional linear and generic nonlinear and/or adaptive black-box, fuzzy, and neural-network models as well as fuzzy-neural models; all seem to have been effective in the respective industrial applications.

Fuzzy [9, 10], neural-network, and fuzzy-neural models [11–14] are known that can be successfully identified from input-output process measurements and yield both quality approximation and good generalization properties. Usually, on-line adaptation of the process model is also needed because of the time-varying behaviour of the furnace, caused by acting disturbances and varying parameters. Notice that fuzzy-neural models combine the advantages of fuzzy and neural-network models as well as possess adaptation features [11, 14, 15].

In real-world constructions of gas/oil fired industrial furnaces, in normal furnace operation, the limits on the controlled temperatures as well as on the actuators' operating control magnitude and slew rate are inevitable. These magnitude constraints give rise to the crucial inequalities to be observed:

$$y_{min} \leq \hat{y}(t+j) \leq y_{max}, \quad j = 1, \ldots, N_p; \tag{1}$$

$$u_{min} \leq u(t+i-1) \leq u_{max}, \quad i = 1, \ldots, N_u; \tag{2}$$

$$\Delta u_{min} \leq \Delta u(t+i-1) \leq \Delta u_{max}, \quad i = 1, \ldots, N_u \tag{3}$$

However, notice that the constraints are imposed on $\hat{y}(t+j)$, that is the jth step ahead prediction of the controlled process variables hence optimizing computation can be employed.

The MPC strategies and system architectures represent some kind of optimum control strategies that are based on employing an explicit model of the plant to be controlled in order to predict reasonably well the future plant responses within a given time horizon and given steps ahead [6, 16–18]. These predictions are then employed to optimize the future behaviour of the plant via adequate synthesis of the control sequence. An optimum sequence of controlling variable(s) is computed by minimizing a chosen cost function, and then the receding horizon strategy is used to implement the optimum sequence. Within this setting the stability issue was resolved. The MPC strategies rely on optimisation of the future process behaviour with respect to future values of the executive control variables, and their extension

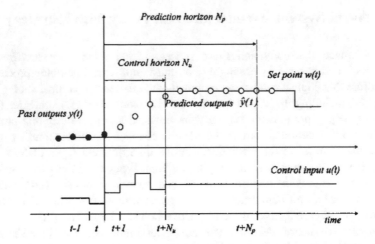

Fig. 1 An illustration of the conceptualization of generalized predictive control theory based on some process model [16]

to use non-linear process models is motivated by improved quality of the prediction of inputs and outputs [19]. In fact, sub-optimum non-linear predictive controls are accomplished this way.

2.1 Sugeno Fuzzy Models

The prediction fuzzy process model can be developed on the grounds of Sugeno or Takagi-Sugeno-Kang (TSK) fuzzy models [20–22]. The TSK fuzzy models (Fig. 2) have been shown to be a suitable representation for rather large class of non-linear systems and to have the potential of universal approximators [11, 13, 15] having wide industrial applications [23]. In addition, they have been shown to be realizable as fuzzy-neural systems [10, 11, 15]. Moreover, also the celebrated adaptive neuro-fuzzy inference system (ANFIS) hybrid algorithm can be used to train such fuzzy-neural representation models [1, 2]. Of course, the choice of the membership function to characterize the respective antecedent fuzzy sets and of the consequent function is to be appropriate to the application domain and the problem considered.

TSK fuzzy models consist of If-Then rules with fuzzy causes in the antecedent and math-analytical functions in the consequent parts, however, these are data driven models. The antecedent fuzzy sets partition the input-signal space into a number of fuzzy regions, whereas functions in the consequent describe system model behaviour within these regions. Given the stochastic features of furnace thermal processes, one should consider a nonlinear input-output model of a system with tuneable parameters in order to arrive at an appropriate process model.

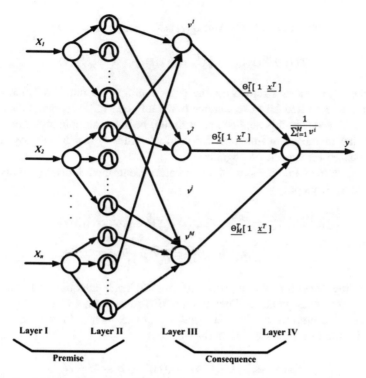

Fig. 2 A signal-flow representation of Sugeno fuzzy-neural models; MISO type [11, 15]

It is therefore that the class of NARX models

$$y(t) = f(x(t)) \qquad (4)$$

is considered along with the following regression vector

$$x(t) = [y(t-1), \ldots, y(t-M); u(t), \ldots, u(t-N+1)]^T \qquad (5)$$

and arbitrary function mapping $f(\cdot)$ defined as appropriate, i.e. *Nonlinear Auto Regressive with eXogenous inputs* (NARX) process models. For cases of when more complex processes, even the *Nonlinear Auto Regressive Moving Average with eXogenous input* (NARMAX) type of system model may be considered, that is the NARX model type extended with a *moving-average* component of the previous predicted errors.

Also, in parallel these consider Sugeno fuzzy models with the proved approximation capacity to approximate arbitrary function $f(\cdot)$. Furthermore, it should be noted that these may be implemented by means of neural networks (see Fig. 2). The rule base of T-S fuzzy system comprises a finite collection of rules:

$$R^{(i)}: \text{If } x_1(t) \text{ is } X_1^{(l)} \text{ And } x_2(t) \text{ is } X_2^{(l)} \text{ And} \dots x_n(t) \text{ is } X_n^{(l)} \qquad (6)$$

$$\text{Then} f^{(l)}(x(t)) = q_0^{(l)} + q_1^{(l)} x(t) + \cdots + q_n^{(l)} x(t)$$

where the upper index l represents the rule identification number. Quantities $X_k^{(l)}$ represents an activated fuzzy set defined in the universe of discourse of the lth input $x_l(t)$. For the needs in this application, it should be noted smooth either bell-shaped or Gaussian membership functions for the fuzzy subsets can be employed due to gradient-based error back propagation.

In fact, due to the physics of temperature controlled reheating process, the simplified TSK model

$$R^{(i)}: \text{If } x_1(t) \text{ is } X_1^{(l)} \text{ And } x_2(t) \text{ is } X_2^{(l)} \text{ And} \dots x_n(t) \text{ is } X_n^{(l)}, \qquad (7)$$

$$\text{Then} f^{(l)}(x(t)) = q_0^{(l)} + q_1^{(l)} x(t)$$

with the linear-function consequent part is sufficiently adequate for this furnace predictive control application. Such a simplified fuzzy model can be regarded as a collection of a number of linear models employed locally in the fuzzy regions defined by the rule premises. Thus the vector

$$x(t) = [x_1(t), x_2(t), \dots, x_n(t)]^T, \quad n = M + N, \qquad (8)$$

defines the subsets of elements in the premise part of the rule l, whereas variables in the lth rule consequent part represent the current values of the fuzzy model input variables. Quantities $q_0^{(l)}$ is a scalar-valued real defining the linear offset, and $q_1^{(l)}$ is a real vector-valued parameter of slopes; both are determined as appropriate in an adaptive manner.

The fuzzifier performs fuzzification of inputs and converts input data from an observed input space into proper linguistic values and the corresponding fuzzy sets through predefined input membership functions. Typically, the input space is represented by the input vector (5) with n elements; the number of the fuzzy subsets for each input variable is chosen to be equal to F. In turn, the number of fuzzy rules is ensured by a combination of n fuzzified input variables and their fuzzy F sets hence it is equal to F^n. It should be noted, to ameliorate rule-ceiling problem, further trade-off between complexity and integrity of this fuzzy-rule-based model is needed [14]. The fuzzy inference engine should match the output of the fuzzifier with the fuzzy logic rules.

With reference to [11], the fuzzy implication in the jth rule (7) and (8) can be realized as follows

$$\mu^{(j)}(x) = \prod_{j=1}^{n} \mu_{X_j^{(l)}}(x_j). \tag{9}$$

In here, $\mu_{X_j^{(l)}}$ represents the membership function of fuzzy subset $X_j^{(l)}$ hence $\mu_{X_j^{(l)}}(x_j)$ specify the fulfilment membership degrees upon the jth fuzzy subset of the corresponding input signal. Solely smooth membership functions can be employed:

(a) Cauchy

$$\mu_{X_j^{(l)}}(x_j) = \left(1 + \left|\frac{x_j - c_{lj}}{\sigma_{lj}}\right|^{2b_{lj}}\right)^{-1} \tag{10a}$$

or Gaussian

$$\mu_{X_j^{(l)}}(x_j) = \exp\left\{-\frac{1}{2}\left(\frac{x_j - c_{lj}}{\sigma_{lj}}\right)^2\right\}, \tag{10b}$$

where c_{lj} represent the centres and σ_{lj} define the widths of these generalized bell-shaped functions for any $b_{lj} > 0$, respectively; and

(b) general sigmoid

$$\mu_{X_j^{(l)}}(x_j) = \left(1 + \exp\{-a_{lj}[x - c_{lj}]\}\right)^{-1}; \tag{11}$$

where c_{lj} represent the potential cross-over points and a_{lj} defines the slopes at $x_j = c_{lj}$.

In the most general MIMO case of fuzzy system models [11, 14] the inference engine for TSK fuzzy models yields

$$y(t) = \frac{\sum_{l=1}^{M} \vartheta^{(l)} y^{(l)}(t)}{\sum_{l=1}^{M} \vartheta^{(l)}}, \tag{12a}$$

$$\vartheta^{(l)} = \mu^{(l)}(x(t)) = \prod_{j=1}^{n} \mu_{X_j^{(l)}}(x_j) \tag{12b}$$

representing the truth value of the lth implication above. Furthermore, it also yields

$$y(t) = \frac{\sum_{l=1}^{M} \vartheta^{(l)} \cdot \theta_l^T \left[1x(t)^T\right]^T}{\sum_{l=1}^{M} \vartheta^{(l)}}, \tag{13a}$$

$$\theta_l^T = \left[q_0^{(l)}, q_1^{(l)}, \ldots, q_n^{(l)} \right]^T. \tag{13b}$$

In the present application to furnace control problem of concern in this work, the above generalized derivations are not needed because the MISO fuzzy model is can serve the purpose of designing the set of temperature predictors for each the three process channels. Thus, for the MISO fuzzy system (6) and (7), after some derivation the following inference formulae are obtained:

$$y(t) = \frac{\sum_{l=1}^{M} f^{(l)}(x(t)) \cdot \mu^{(l)}(x(t))}{\sum_{l=1}^{M} \mu^{(l)}(x(t))}, \tag{14}$$

and

$$y(t) = \sum_{l=1}^{M} f^{(l)}(x(t)) \cdot \vartheta^{(l)}(x(t)) \tag{15}$$

with

$$\vartheta^{(l)}(x(t)) = \frac{\prod_{j=1}^{n} \mu_{X_j^{(l)}}(x_j)}{\sum_{l=1}^{M} \mu^{(l)}(x(t))} \tag{16}$$

the normalized value of the membership function degree on the arbitrary l-th acting fuzzy- rule.

The actual implementations of the relevant fuzzy predictors in this research have been obtained by appropriately shifting the inputs of the fuzzy based model as inferred from Eqs. (4) to (5) and on the grounds of TSK fuzzy model of the plant. Therefore a sequential algorithm, based on the knowledge of current values of the regression vector $x(t)$, along with the fuzzy inference (15), computes

$$y(t+k-1) = \sum_{l=1}^{M} f^{(l)}(x(t+k-1)) \cdot \vartheta^{(l)}(x(t+k-1)) \tag{17}$$

the arbitrary kth step ahead predicted process output, and the needed subsequent regression vector

$$x(t+k) = [y(t+k-1), y(t+k-2), \ldots, y(t), \ldots, u(t+k), u(t+k-1), \ldots, u(t)]^T \tag{18}$$

Further, the complementing procedure section of the identification involves structure identification and estimation of the unknown parameters [12, 20–22]. The structure of the fuzzy model involves the number of membership functions for each input and their shape is determined by the unknown parameters in the premise part, e.g., say (c_{lj}, σ_{lj}), of each rule. For the simplified TSK model (7) and (8), the unknown parameters in the consequent part are represented by the coefficients in quantities $q_0^{(l)}$ and $q_0^{(l)}$. These are determined so as to achieve the best feasible fitting on the input-output data, and any efficient procedure can be used [12].

Within computer simulation platform, one apparent possibility is to use stochastic disturbance excitation conveniently implemented as additive Pseudo-Random Binary Sequence (PRBS). The results of simulation under PI control and using a linear time-delay model at 80 % operating power enabled obtaining data sets in a sequence of 60,000 samples for each of the furnace input-output temperature channels. Thus formed data set for each channel is employed as follows: (i) the first 43,000 samples are used to train a TSK neuro-fuzzy adaptive model of the system to-be-controlled. The remaining 17,000 samples are used in terms of validation data set.

It was found via test simulation that the regression vector (5) can be defined with the parameters $M = 2$ and $N = 2$. The chosen fuzzy model to be identified is the Gaussian one with five fuzzy inputs; they are the regression vector elements and one to the operating load characterizing the output. The number of fuzzy subsets (membership functions) assigned to the first 4 inputs of the fuzzy model was set to two, while those assigned to the operating load characterization was set to five. Thus, the resulting number of rules in the knowledge base of $2^4 \times 5 = 80$ remained within a manageable size for the fuzzy-neural model.

The fuzzy knowledge base of 80 rules needed: in the premise part, the 13 Gaussian membership functions require 26 parameters; and the linear functions in the consequent part require 6560 coefficients. Thus the identified fuzzy-neural model needs automated training and was trained by employing ANFIS hybrid algorithm and the respective recommendations [11] along with using MATLAB Fuzzy Logic Tool-Box; once trained it was used as predictor in this predictive control strategy. As it is well known, ANFIS hybrid algorithm exploits the mismatch between model output and the output in training data-set in order to adjust the parameters of the consequent, $(a^{(l)}, b^{(l)})$, and of the premise, (c_{lj}, b_{lj}). Jang's adaptation algorithm makes use of least squares estimates of the consequent parameters, and the gradient-based error back propagation for adapting the premise parameters. Typically, on a standard PC platform Pentium IV, 2.8 GHz CPU, 512 MB RAM, 80 GB HDD, the adaptation of the developed fuzzy-neural model takes about 15 min/epoch for a data-set of 43,000 entries using MATLAB Fuzzy Logic Toolbox [24].

2.2 Design of the Fuzzy Model-Based Predictive Control Strategy

As the general MPC strategy is based on three main concepts, the first one being use of a model to predict the system output at the future discrete time instants over a defined prediction horizon N_p, the second one—computation of the future control actions over a control horizon $N_u \leq N_p$ at the same time instants via minimizing some objective function, while the third one is the receding horizon strategy the choice of these is needed. Only the first control action in the sequence is applied, the horizons are moved towards the future and optimization procedure is repeated. The control variable is manipulated only within the control horizon and it remains constant afterwards.

The predicted output $\hat{y}(t+k\,|\,t)$, $k = 1, 2, \ldots, N_p$, depends on the input and output values up to the time t and the respective future control signals $u(t+k|t)$, and these are the values to be calculated. Since the objective is to keep the output $y(t+k)$ as close as possible to a desired reference $r(t+k)$, which in furnace control is the operating equilibrium set-point command, the simplest objective function is defined by the following quadratic form:

$$ J(t, U) = \sum_{k=1}^{N_p} [r(t+k) - y(t+k)]^2 + \beta \sum_{k=1}^{N_u} [\Delta u(t+k-1)]^2, \quad \beta > 0, \qquad (19) $$

where U represents the control sequence. Notice it employs the instantaneous incremental control action $\Delta u(t) = u(t) - u(t-1)$ at t that is any arbitrary time instance. Once the optimal sequence is calculated only the signal $u(t|t)$ is implemented; for the time $t+1$ a new output value $y(t+1)$ is measured and a new control sequence is calculated.

The here developed predictive control system exploits most of the ideas of the generalized predictive control strategy, and it assumes the plant output equation

$$ y(t) = \hat{y}(t) + \eta(t). \qquad (20) $$

It is consisted of $\hat{y}(t)$ that is generated by the fuzzy-neural model (14–16) and $\eta(t)$ that is generated by a suitable noise model. Thus the main tuning parameters are the prediction horizon, N_p, the control horizon, N_u, and the weighting factor β penalizing changes in the control actions given the limits on control (2) and its slew-rate (3). It should be noted, a long control horizon allows a more active control values, and it is thus likely to enable better performance. On the other hand, a short horizon generally makes the control system more robust [19].

With the quadratic objective function and assumption of no constrains on the control action, cost function can be minimized analytically. The latter is justified on the grounds of using an appropriate penalizing factor and a fuzzy-neural predictor that is essentially nonlinear. Objective function J is minimized with respect to the sequence of control actions via

$$ \partial J / \partial u = 0. \qquad (21) $$

In vector notation, quadratic objective function (19) can be re-written as follows:

$$J(t, U(t)) = E(t)^T E(t) + \beta \Delta U(t)^T \Delta U(t) = J(t, \Delta U(t)), \tag{22}$$

$$E(t) = R(t) - Y(t), \tag{22a}$$

$$R(t) = \left[r(t+1), r(t+2), \ldots, r(t+N_p) \right]^T, \tag{22b}$$

$$Y(t) = \left[y(t+1), y(t+2), \ldots, y(t+N_p) \right]^T, \tag{22c}$$

$$\Delta U(t) = \left[\Delta u(t+1), \Delta u(t+2), \ldots, \Delta u(t+N_u - 1) \right]^T \tag{22d}$$

with $\Delta u(t+1) = u(t+1) - u(t)$ etc., and also

$$U(t) = \left[u(t+1), u(t+2), \ldots, u(t+N_u - 1) \right]^T, \tag{22e}$$

of course. Thus calculation of the gradient vector of the cost function (22) at the moment t with respects to predicted control actions yields the minimization:

$$\nabla J(t, U) = \frac{\partial J(t, U)}{\partial U} = \left[\frac{\partial J}{\partial u(t)}, \frac{\partial J}{\partial u(t+1)}, \ldots, \frac{\partial J}{\partial u(t+N_u - 1)} \right]^T. \tag{23}$$

Observe Eq. (22) and notice that for the first element one can obtain the following matrix equation:

$$\nabla J(t, U) = -2E(t)^T \frac{\partial Y(t)}{\partial U} + 2\beta \Delta U(t) \frac{\partial \Delta U(t)}{\partial U}. \tag{24}$$

Finally, given $\nabla J(t, U) = 0$, after a lengthy analysis of the partial derivatives in (24), finally the system of equations computing the optimal values for the incremental control actions appeared to be given by

$$\Delta U(t) \frac{\partial \Delta U(t)}{\partial U} = \beta^{-1} E(t)^T \frac{\partial Y(t)}{\partial U} \tag{25}$$

yielding

$$\Delta u(t+N_u - 1) = \frac{1}{\beta} \sum_{k=1}^{N_p} e(t+k) \frac{\partial y(t+k)}{\partial u(t+N_u - 1)}, \tag{26a}$$

$$\Delta u(t+N_u - 2) = \Delta u(t+N_u - 1) + \frac{1}{\beta} \sum_{k=1}^{N_p} e(t+k) \frac{\partial y(t+k)}{\partial u(t+N_u - 2)}, \tag{26b}$$

$$\Delta u(t) = \Delta u(t+1) + \frac{1}{\beta} \sum_{k=1}^{N_p} e(t+k) \frac{\partial y(t+k)}{\partial u(t)}. \tag{26c}$$

Equation (26) represent a computationally effective iterative optimization algorithm. At the next time instant $t + 1$, all the tune depended quantities in (26) are to
be shifted by one step, and the entire optimization routine is started and carried out
again. And the same can be carried out until all the needed computations be
accomplished. And this completes the fuzzy MBCP strategy for steel-mill pusher
furnace control and high-power industrial furnaces alike proposed in here.

3   Simulation Results for RZS Furnace

In the previous redesign case-studies on RZS furnace at Skopje Steelworks the
results on the operating steady-state characteristics as well as the respective schematic diagram have appeared [1, 2]. The schematic is presented in Fig. 3 and
layouts the basic element of the furnace. The furnace, which has three zones, a total
size of $25 \times 12 \times 8$ m, maximum installed 28 MW and normal operating 25 MW
powers, is operated in a steady-state regime at a given pusher pace depending on
slab/ingot size and other metallurgical specifications. The standard slab heat processing is carried out at 1050–1250 °C and the pushing pace 5–9 min per unit slab.
The pusher pace governs steel-mill furnace operation. The normal operating
regimes are presented in Figs. 4 and 5 respectively. On Fig. 4 we can notice the
nonlinearity of the characteristics of the furnace in the region of pushing pace of 5–
7 min.

Locally at the nominal operating equilibrium, the linearized thermal process
dynamics ($K = \Delta\Theta/\Delta Q$) in each of the furnace process channel can be represented
as

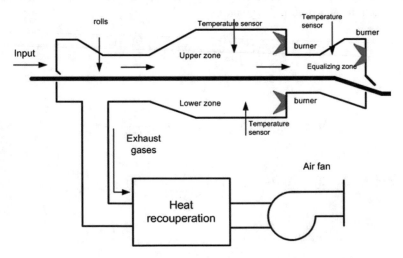

Fig. 3 Schematic of the three zone 28 MW industrial furnace in Skopje Steelworks

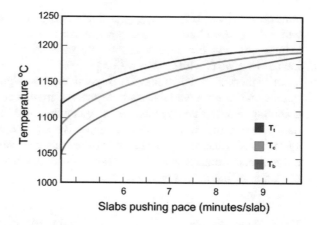

Fig. 4 Normal operating regime T_{zone} = 1150 °C and 90 % power: 10 % change in controlled temperature at the upper heating zone (initial condition and desired reference temperature $y_{SS}(0)$ = 1150 °C)

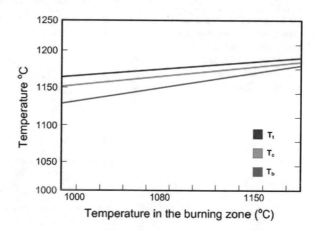

Fig. 5 Normal operating regime T_{zone} = 1150 °C and 90 % power: 10 % change in controlled temperature at the equalizing zone (initial condition desired reference temperature $y_{SS}(0)$ = 1150 °C)

$$G_{ij}(s) = \frac{K_{ij}\exp(-\tau_{ij}s)}{(T_1+1)(T_2s+1)} \quad (27)$$

with certain ranges of values for all parameters in process channels (i, j), i = 1, 2, 3 and j = 1, 2, 3. Besides, the ranges of value for the main attributes of slab heating dynamics have been identified: steady-state (SS) gains K_{ij} (in [°C/MJ/min]) and pure time-delays τ_{ij} (in [min]) within each of the process signal transfer paths (i, j); these enabled to estimate an average pure time delay τ_{av} in the furnace temperature process dynamics. The identified numerical values are found in [1, 2].

The residing horizon for the predictive control can be determined by observing the operating pure time delay phenomena and the relationship between pure delay and inertial times $\delta_1 = \tau_{ij}/(\tau_{ij} + T_{ij})$ and $\delta_2 = \tau_{ij}/T_{ij}$, taking into consideration the adopted sampling period T_s = 1 [min] and the integer part of $d = [\tau_{ij}/\tau_{ij}T_s \cdot T_s]$. It should be noted, the $y_i(0)$ = 1150 °C implies non-zero initial values of actuator control variables $u_i(0) = U_{midrange}$ representing the mid-range control values to which fuzzy MPC controls are superimposed. The process noise vector $\eta(t)$ was assumed to have the same stochastic characteristics in every channel and approximated as a weak white noise with $\sigma_\eta^2 = 10$.

In Figs. 6, 7, 8, 9, 10 and 11, respectively, the signal values of the respective output and control variables are expressed in percentages (%) of the nominal operating equilibrium values for temperatures and the corresponding controls manipulating fuel inflows for the respective furnace zones. The overall control system performance is inferred from these sets of sample results on controlled outputs and control sequences. These guarantee reasonably smooth optimized operation of the reheating furnace.

Fig. 6 Normal operating regime $T_{zone} = 1150$ °C and 90 % power: 10 % change in controlled temperature at the lower heating zone (initial condition and desired reference temperature $y_{SS}(0) = 1150$ °C)

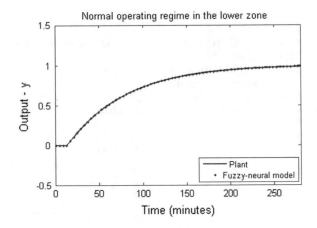

Fig. 7 Normal operating regime $T_{zone} = 1150$ °C and 90 % power: 10 % change in controlled temperature at the upper heating zone (initial condition and desired reference temperature $y_{SS}(0) = 1150$ °C)

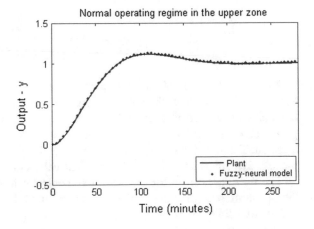

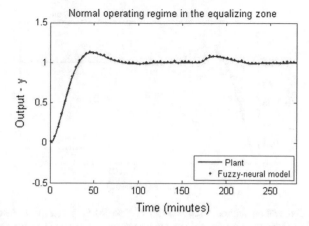

Fig. 8 Normal operating regime T_{zone} = 1150 °C and 90 % power: 10 % change in controlled temperature at the equalizing zone (initial condition desired reference temperature $y_{SS}(0)$ = 1150 °C)

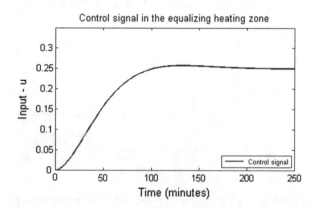

Fig. 9 Normal operating regime T_{zone} = 1150 °C and 90 % power: time history of control changes for 10 % increase in controlled temperature at the lower heating zone (initial condition $u_i(0) = U_{midrange}$)

4 Discussion and Concluding Remarks

In this paper, a fuzzy predictive control design for the high-power pusher-type furnace temperature control has been presented. This control system employs a fuzzy-neural model to implement the predicting function and a gradient-optimization algorithm to synthesize the controlling sequence and close the control loop. Thus it represents a hybrid soft-computing and math-analytical control strategy. Sufficiently many inputs, characterized by fuzzy subsets (the regression vector and the current output) were employed in order to achieve

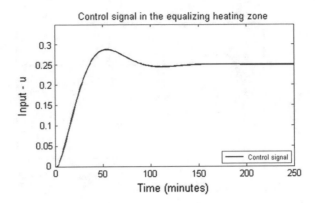

Fig. 10 Normal operating regime T_{zone} = 1150 °C and 90 % power: time history of control changes for 10 % increase in controlled temperature at the upper heating zone (initial condition $u_i(0) = U_{midrange}$)

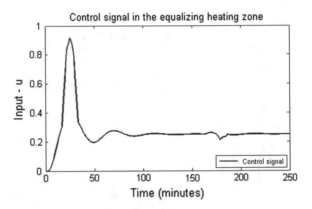

Fig. 11 Normal operating regime T_{zone} = 1150 °C and 90 % power: time history of control changes for 10 % increase in controlled temperature at the equalizing zone (initial condition $u_i(0) = U_{midrange}$)

versatility of the fuzzy-neural model. On the other hand, the least number of two membership functions have been employed for the sake of rule-base compactness. The main advantage of the proposed fuzzy MPC lies in the fact that it implements generically non-linear model-based predictive control. The easy fuzzy model description and the easy computation of the gradient vector during the optimization procedure appear to be the main advantages of the computation algorithms. For practical system engineering and maintenance reasons, the digital implementations are sought within a standard computer process control platform.

A promising future research is towards empowering further the present design by introducing an expert subsystem to accommodate fuzzified goal-reference and operating constraints to the temperature outputs. Another direction is to elaborate the synthesis methodology exploiting math-analytical optimization solutions and their combination to construct a non-linear overall model predictive control, departing operational composition of several locally linear model predictive control strategies, so as to avoid the time consuming tune search procedures.

Acknowledgments The Authors gratefully acknowledge the crucial contribution by Prof. G.M. Dimirovski in proving the theoretical results reported in this article.

References

1. G.M. Dimirovski et al., Learning control of thermal systems, in *Control of Complex Systems* eds. by K.J. Astroem, P. Albertos, M. Blanke, A. Isidori, W. Schaufelberger, R. Sanz (Springer, London, 2001), pp. 317–337
2. G.M. Dimirovski, A.T. Dinibutun, M. Vukobratovic, J. Zhao, Optimizing supervision and control for industrial furnaces: predictive control based design (Invited Plenary Lecture), in *Automatic Systems for Building the Infrastructure in Developing Countries—Global and Regional Aspects DECOM-04*, eds. by V. Sgurev, G. Dimirovski, M. Hadjiski, (Union for Automation & Informatics of Bulgaria and the IFAC, Sofia, BG, 2004), pp. 17–28
3. J. Rhine, R. Tucker, *Modelling of Gas-fired Furnaces and Boilers* (McGraw-Hill, New York, 1991)
4. Y.Z. Lu, T.J. Williams, Computer control strategies of optimal state feedback methods for the control of steel mill soaking pits. ISS Trans. **2**, 35–43 (1983)
5. Y.Y. Yang, Y.Z. Lu, Dynamic model based optimization control for slab reheating furnaces. Comput. Ind. **10**, 11–20 (1988)
6. H. Demircioglu, D. Clarke, CGPC with guaranteed stability properties. IEE Proc. Pt. D Control Theory Appl. **139**(4), 371–380 (1992)
7. Z. Icev, M. Stankovski, T. Kolemisevska-Gugulovska, J. Zhao, G. Dimirovski, Pusher Reheating Control: A Fuzzy-neural Model Predictive Strategy, in *Proccedings of the IFAC WS on Energy Saving Control, Bansko, BG*, October 2–4, paper. (The IFAC and Bulgarian Union for Automation & Informatics, Sofia, BG, 2006), pp. 33–38
8. E. Camacho, C. Bordons, *Model Predictive Control* (Springer, London, 2004)
9. L.A. Zadeh. Outline of a new approach to the analysis of complex systems and decision processes. IEEE Trans. Syst. Man Cybern. **1**(1), 28–44 (1973)
10. L.A. Zadeh, Fuzzy logic, neural networks, and soft computing. Commun. ACM **37**(3), 77–84 (1994)
11. J.-S.R. Jang, C.-T. Sun, E. Mizutani, *Neuro-Fuzzy and Soft Computing* (Prentice Hall, Upper Saddle River, NJ, 1997)
12. L. Ljung, *System Identification: Theory for the User*, 2nd edn. (Prentice Hall, Upper Saddle River, NJ, 1999)
13. L.X. Wang, *Adaptive Fuzzy Systems and Control: Design and Stability Analysis* (Prentice Hall, Englewood Cliffs, NJ, 1994)
14. G.M. Dimirovski, Complexity versus integrity solution in adaptive fuzzy-neural inference models. Intl. J. of Intell. Syst. **23**(5), 556–573 (2008)
15. C.-T. Lin, C.S.G. Lee, *Neural Fuzzy Systems* (Prentice Hall, Upper Saddle River, NJ, 1996)
16. D.W. Clarke, C. Mohtadi, P.S. Tuffs, Generalised predictive control: Pt. 1, the basic algorithm; Pt. 2, extensions and interpretations. Automatica **23**(2), 137–160 (1987)
17. D.W. Clarke, C. Mohtadi, Properties of generalized predictive control. Automatica **25**(6), 859–876 (1989)
18. M. Morari, J.H. Lee, Model predictive control: past, present, and future. Comput. Chem. Eng. **23**, 667–682 (1999)

19. F. Allgower, T. Badgwell, J. Qin, J. Rawlings, S. Wright. Nonlinear predictive control and moving horizon estimation—an introductory overview, in *Advances in Control: Highlights of ECC'99*, ed. by P.M. Frank. (Springer, London, 1999), pp. 391–449
20. T. Takagi, M. Sugeno, Fuzzy identification of systems and its applications to modelling and control. IEEE Trans. Syst. Man Cybern. **15**(1), 116–132 (1985)
21. M. Sugeno, G.T. Kang, Structure identification of fuzzy model. Fuzzy Sets Syst. **28**, 15–33 (1988)
22. M. Sugeno, T. Yasukawa, A fuzzy-logic based approach to qualitative modelling. IEEE Trans. Fuzzy Sets Syst. **1**(1), 7–31 (1993)
23. K. Hirota (ed.), *Industrial applications of fuzzy technology* (Springer, New York, 1993)
24. Mathworks. Matlab/Simulink: Version 5.3—Release R11. The Mathworks, Boston, MA (1999)

Exactus Expert—Search and Analytical Engine for Research and Development Support

Gennady Osipov, Ivan Smirnov, Ilya Tikhomirov, Ilya Sochenkov
and Artem Shelmanov

Abstract The paper presents the system-"Exactus Expert"—search and analytical engine. The system aims to provide comprehensive tools for analysis of large-scale collections of scientific documents for experts and researchers. The system challenges many tasks, among them full-text search, search for similar documents, automatic quality assessment, term and definition extraction, results extraction and comparison, detection of scientific directions and analysis of references. These features help to aggregate information about different sides of scientific activity and can be useful for evaluation of research projects and groups. The paper discusses general architecture of the system, implemented methods of scientific publication analysis and some experimental results.

1 Introduction

Tasks of R&D activities management become more and more important nowadays. Development of both scientific and economic potentials of states depends on it. Expert evaluation of research projects submitted to various calls and grant applications is a notable part of a decision making process in R&D. Such projects should be innovative, urgent, and their results are supposed to be of high demand at the market. It also applies to search for perspective innovative projects, search for the points of growth in the research community. These tasks are relevant both for the state authorities that are responsible for creation of the state policy in the research field and for private investors.

G. Osipov · I. Smirnov · I. Tikhomirov (✉) · I. Sochenkov · A. Shelmanov
Institute for Systems Analysis of the Russian Academy of Sciences,
Prospekt 60 Let Oktyabrya 9, 117312 Moscow, Russia
e-mail: tih@isa.ru

G. Osipov
e-mail: gos@isa.ru

© Springer International Publishing Switzerland 2016
M. Hadjiski et al. (eds.), *Novel Applications of Intelligent Systems*,
Studies in Computational Intelligence 586, DOI 10.1007/978-3-319-14194-7_14

269

Progress of science and its integration with IT leads to the fact that the tasks related to exploration of state of the art in various research areas become more and more significant. It is important to have a clear view of what scientific fields and research topics are intensively developing at the present time, what directions tend to decline and what directions are expected to advance in the near future.

A scientific publication normally gives essential information about research including problem statement, proposed solutions and achieved results. Scientific publications play the vital role in presentation of research. A large amount of information depicting the current state of science and technology in the world is now freely available on the web. Such information can be found in electronic versions of scientific and popular journals, preprints, reports on R&D works presenting research results, their potential economic impact and recommendations on their use and applications.

To make use of this vital information one requires complex analytical tool that can process large amounts of unstructured and semistructured data, which is basically represented by texts in natural languages. This paper presents the system "Exactus Expert"—search and analytical engine that aims to provide complex analysis of large-scale collections of scientific documents and support R&D activities.

2 Related Work

There are many systems that provide analytical support of scientific activities. However, most of them are primarily focused on analysis of bibliographic references of scientific publications. Just few systems provide abilities for deep comprehensive analysis of scientific trends and domains.

SciVerse Scopus is positioned as the biggest universal referential database of scientific information. It covers papers from more than 5000 international publishers including Russian journals. In addition to advanced search ability and analysis of references, the system provides tools for comparison of journals by publication activity and by other metrics [10].

SciVal is a complex system for branch of science analysis. It allows evaluating scientific results of different branches and provides graphical visualization of effectiveness of organizations, countries and geographical regions for a period of time and domain (map of science) [1].

Web of Knowledge is a system that provides access to the most well-known citation index Web of Science that contains a wide range of publications in almost every branch of science. There are advanced customizable search abilities and tools for analysis of bibliographic references. The system provides comprehensive statistics about journals and publication for different domains. There are also some tools for monitoring of scientific trends and research teams [7].

Although every system has a lot of abilities and tools, they are not universal but better suit particular cases. The tasks of scientometrics i.e. determining trends, evaluating research teams and their results, finding novel outstanding approaches is still far away from being solved. Another drawback of state of the art systems is that they mostly deal with structured data, which is prepared manually. However, most scientific documents available on the web are unstructured. There is a lack of systems that can deal with raw unstructured data like texts in natural languages. Although there are some examples like Google Scholar that process raw data, these systems provide lesser abilities for analytics of science.

We propose the "Exactus Expert" system which challenges tasks of processing raw unstructured data like texts in natural language and provides tools for comprehensive analysis of scientific papers, domains and research groups.

3   Architecture of "Exactus Expert"

In general, "Exactus Expert" works as traditional search engine. It downloads papers and other documents from the web, performs linguistic analysis and other processing on them and store extracted data in semantic index and database. Then the extracted and stored information is used by search and analytical modules. We have developed our own crawling, storage and search systems, as well as analytical modules. All of them are designed to be distributed, so it is possible to scale system to multiple servers. General architecture of the system is represented in Fig. 1.

3.1   Crawling and Initial Data

As initial data, we use scientific publications from open sources available on the web—websites of research and popular journals, websites of scientific conferences, preprints and other sources. The crawling system searches for publications,

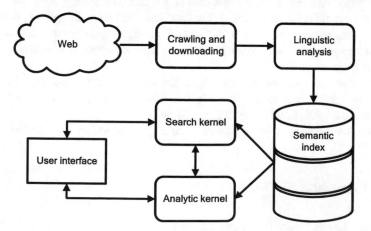

Fig. 1 General architecture of "Exactus Expert"

downloads them and performs some preprocessing on acquired documents. It purifies formatted text of the document from the rest of the webpage and extracts from the webpage some metadata of the document: author(s), title, date of publication etc. The crawling system converts acquired documents to the internal format from origin format. It supports "html", "pdf", "doc", "docx", "odt" and other widely used text formats. There is also ability to automatically detect ill-formed documents ("pdf" and "ps"), which text could not be extracted by the ordinal procedures. For these cases, open-source OCR tools are integrated into the crawling system of "Exactus Expert" to recognize texts from images. The system also groups acquired documents in collections belonging to the same topic and origin. Documents are divided into the following groups according to research area:

- physics and mathematics;
- engineering;
- chemistry;
- agriculture;
- biology;
- arts and humanities;
- socio-economic studies;
- medicine;
- geosciences.

Current experimental database includes about 11 million documents in Russian and English. It includes journal publications, theses abstracts, conference papers and patents.

The crawling system is driven by set of handcrafted "configurations", which are written in some sort of declarative language. They contain patterns for searching scientific publications within the web source, rules for text and metadata extraction. The "configurations" also include information about predefined topic and language of the web source.

The major problem of initial data crawling is that web sources are heterogeneous and have complicated structure. Moreover, structure of webpages changes from time to time, making the current versions of "configurations" obsolete. This makes hard to engage, tune and support big amount of different web sources, since it requires a lot of handwork.

There are some known methods for automatic webpage crawling and metadata extraction based on machine learning [5]. We are looking forward to involving them in next versions of the crawling system.

After downloading, every acquired document is subjected to linguistic analysis.

3.2 Linguistic Analysis

"Exactus Expert" performs complex and comprehensive linguistic analysis of natural language texts in Russian and English. The main stages of its natural language processing pipeline are graphematic, morphological, syntactic, semantic analysis and coreference resolution. For basic graphematic, morphological, and syntactic analysis we use open source frameworks: AOT for Russian [24] and Freeling for the English language [18]. For semantic analysis and coreference resolution we use our own framework.

The side frameworks produce clauses (in natural language clauses correspond to simple sentences in a compound sentence, participial and adverbial participial phrases and other constructions) and syntax relations. Then they are converted into special syntax dependency trees. There are two types of syntax trees: trees built in a clause connect words and trees built in a sentence connect clauses. Morphological features of words are enriched with categorial semantic class (CSC). Categorial semantics is a generalized meaning characterizing words that belong to the same categorial class (for instance, nouns may belong to the classes of people, things and attributes) [16]. CSC is necessary for further semantic analysis, since it defines the syntax features of the word and ways of functioning in clause. Identification of CSC of words is based on other morphological properties of words, special dictionaries and some heuristic rules. The result syntactic structure of text is passed to semantic analysis.

The semantics analysis implemented in the system is based on methods and models of the relational-situational analysis [14, 16, 22], which in their turn rely on the theory of linguistic semantics [29, 30]. The underlying semantic model of text is heterogeneous semantic network [13, 22]. The task of semantic analysis is in some way similar to the semantic role labeling task. It consists of revealing semantic meanings of nominal syntaxemes and establishing semantic relations on the set of syntaxemes. Syntaxemes are minimal indivisible semantic-syntactic structures of language [12]. They are characterized by:

- categorial semantics of the word (CSC);
- morphological form;
- function in the sentence.

Semantic relations between syntaxemes express relations between concepts in the conceptual system of the domain. The procedure of semantic analysis rely on predicate words of sentences, predicate word dictionary and machine learning [15]. When the target clause contains predicate verb, roles and relations are assigned to syntaxemes accordingly to predicate word dictionary. Predicate word dictionary contains information about possible meanings, relations, feature sets that syntaxemes must have to acquire particular meanings and compatibility of relations with meanings. The list of predicate words, say, for the Russian language contains verbs and various verb derivative forms. The dictionary was developed manually by linguists and contains 95 % of mostly frequent Russian verbs. There is also version of the dictionary for English. When it is not possible to determine predicate word for clause, the procedure based on logical machine learning method assigns meanings to syntaxemes accordingly to their contexts. The result structure built by semantic analysis is semantic network of sentence.

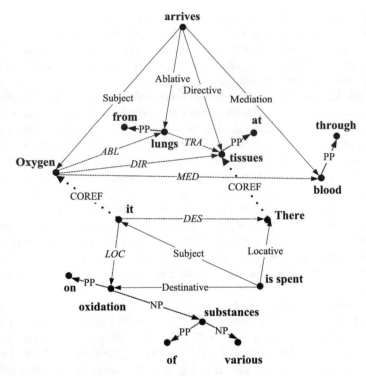

Fig. 2 Visual representation of semantic network of text: "Oxygen arrives at tissues from lungs through blood. There it is spent on oxidation of various substances"

Coreference and anaphora resolution links separate semantic networks of sentences in the semantic network of whole text. These relations are established using lexical databases like WordNet and following features: distance, morphological properties and syntax role. There are also some predefined heuristic rules.

Semantic networks of texts generated as the result of the linguistics analysis are stored in semantic indexes and are used for search and further high-level analysis. Figure 2 shows visual representation of sample semantic network.

The complex linguistic analysis is slower than methods that are used in traditional search engines. However, complex natural language processing gives advantages in deep analysis of documents.

3.3 Storage System

Storage system consists of a database for metadata and a set of indexes used to provide different functionality. The first one is common relational database and mostly perform service functions. The storage system contains semantic inverted index, inverted spectral index, and also some service indexes. Semantic inverted index provides the ability of full-text search through documents and their metadata like title, authors, date of publication and others. Inverted spectral index provides functionality for fast similar document search [4].

The remarkable thing about semantic inverted index is that it stores not only information about word occurrences in documents and text formatting as well as traditional search engines do, but also stores almost all information from semantic network of text generated by linguistic analysis. Semantic index is also able to store additional metadata about text regions via standalone markup. This allows searching only through text regions with special mark.

Storing comprehensive information about text allows building intelligent ranking functions and implement such features as question answering search. This information also becomes available to complex high-level analytical modules that need deep understanding of text.

Although we store a lot of additional information in semantic index like syntax/ semantic relations, semantic meanings, morphological information etc. However, our technology of inverted index organization allows effectively storing full-text indexes. The average overhead on additional information do not exceed 16 % compared to simple keyword index stored by traditional way of packing [17]. The traditional inverted index data structures show average overhead of 68 % for full-text index with additional information, which is significantly worse than the data structures we proposed and implemented.

Performance of the storage system is about 28 MB per minute measured on a single middle-class server node (to speed up it is possible to launch instances on many nodes). Configuration of the node was Intel Core i7 CPU 3.1 GHz, 16 GB RAM, 4xHDD—500 GB 7200 rpm, RAID 1. For example, the performance of the storage system allows indexing Russian Wikipedia (960 thousand documents) in 6.5 h.

4 Technologies for Analysis of Scientific Publications

The search and analytical engine "Exactus Expert" provides many powerful features of text processing and scientific publication analysis. The most important features are briefly described in this section.

4.1 Semantic Search

The primary feature of "Exactus Expert" system is semantic search technology [14, 16, 23]. As well as the simple keyword search, the system can perform semantic search when it looks for the documents that suit request by meaning. There is ability to refine query by providing some metadata like authors, year of publication or title. The system's query language allows searching particular phrase when user needs documents that contain particular noun or verb phrases. It is possible to restrict set of words to be in a single sentence.

To achieve all of these features the ranking algorithm takes into account complex information from semantic index. The main idea of semantic search is comparing semantic networks of query and documents stored in semantic index. Semantic networks are compared in words, semantic meanings and semantic relations [13]. Search algorithm also uses stored syntax relations to rank matches of syntax structure higher. In addition, ranking algorithm also takes into account morphological properties and text formatting markup.

To find documents with exact phrase the system searches only for documents containing phrases, that are syntactically similar to the query. So in this mode, when user searches for "*Regional library*" he may find "*Regional public libraries*" but not "*Regional problems of libraries*".

Since metadata of documents is also stored directly in semantic index in the same way as formatting markup it is possible to perform full-text search through the metadata.

On the basis of the described approach system implements advanced features that allow to compare documents with collections and collections with collections using complex linguistics information [23].

4.2 Search for Similar Documents

The considered task consists in finding documents that are close to the given document by theme. The main idea of the proposed method for similar document search is comparing sets of top most significant keywords and key phrases of documents. As key phrases, we use syntactically related words that are not considered as stop words.

Selection of the most important keywords and key phrases that provide best description of the document is not a trivial task. The significance of keyword or key phrase is described by its weight, which can calculated in many ways. We consider traditional *TFIDF* value [6, 9] (term frequency (*TF*) multiplied by inverse document frequency (*IDF*)) not enough reliable for that purpose and propose modified version of that value [8]. Instead of term frequency, we use *lTF* value, which is defined as follows:

$$lTF(w) = \log_s(k(w) + 1) \tag{1}$$

w—keyword or key phrase; S—total formatting markup weight of all words in the document; k—total formatting markup weight of all occurrences of the keyword or key phrase in the document.

lTF is better than *TF*, since it is less sensitive to document size. It also takes into account text formatting markup.

Similarity between two documents is computed as distance between two vectors composed of weights of keywords and key phrases of those documents. For this purpose, we use Hemming distance measure as well as Cosine distance and their combination.

To speed up procedure of keyword and key phrase set comparison inverted spectral index was implemented [4].

4.3 Retrieving Results from Papers

Results of research presented in a paper are formulated by means of specific phrases, which correspond to special structures. Theses structures contain pairs $\langle predicateword, meaning \rangle$ of special predicate word and meaning of its argument (syntaxeme).

For extraction of such structures a corpus of scientific texts with marked up phrases describing results was formed. Using Bayesian classifier structures for extracting results were obtained. For example, a result can be presented with the structure $\langle develop, object \rangle$, so the sentence *"Authors developed the method"* is considered describing result.

It was discovered that theoretical results are commonly presented with structures $\langle predicateword, delibirative \rangle$, and applied results are presented with structures $\langle predicateword, object \rangle$.

Retrieving results allows evaluating efficiency of a given research or a given field of research and makes possible to compare them by productivity.

4.4 Term Definition Extraction

Definitions of terms in scientific publications can be determined by their lexical, syntactic and semantic contexts. To find and extract term definitions we developed method based on analysis of these contexts [19]. Our linguists revealed more than 60 contexts of term definitions. They were refined and generalized during some experiments. In the result, we created set of syntactic and semantic templates that cover the most frequent cases of term definitions. The idea of the method is to search all matches that suit lexical, syntactic and semantic conditions of stored templates. For example, template *POS(Noun) & SemValue(Estimative) + Lemma ("called")* matches definition *"This dividing line is called the bissectrice or bisection line"*. Fifteen templates of such kind were implemented.

Templates also bear information about, which part of the match should be extracted as a term and which part should be treated as a definition. When list of terms is constructed, it is filtered using some heuristic rules. These rules exclude from the result set typical erroneous definitions and redundant words from terms themselves.

The set of found terms of a document can extend a list of keywords, it can be taken into account by procedure of automatic annotation construction and it can be a sign of novelty of scientific paper. Terms and definitions (as well as results) can also obtain a special mark in a search index, which has influence on relevance of the document to a query. Terms and definitions placed into the search index can help to trace relationships between documents since it is possible to find texts with similar terminology and even determine in what text term was introduced first.

4.5 Assessment of Quality of Scientific Publications

The problem of evaluating quality of scientific publication has two aspects. First, the publication should have conventional format, i.e. contain sections such as problem statement, methods, solutions, results of the research, conclusions, references and so on (see, for example, IMRAD [2, 25]). Second, the publication should not contain quasi-scientific non-scientific lexis and phrases.

To check the paper structure it is necessary to detect availability of mentioned sections [20]. As for retrieving results, we assume that sections contain specified semantic structures such as ⟨predicateword, argument, delibirative⟩. A corpus of scientific texts with marked up sections was created. Bayesian method extracted structures specific to the sections. Thus, the section "problem statement" frequently contains structures ⟨is, research, object⟩, ⟨attract, attention, subject⟩ etc., the section "conclusion" contains structures ⟨discover, opportunities, resultative⟩, ⟨present, we, subject⟩, ⟨let, research, causative⟩ etc.

For checking quasi-scientific or prescientific lexis and phrases in publications, the special dictionaries were developed.

4.6 Analysis of References

"Exactus Expert" implements common research networking tools' features. The system automatically extracts authors and bibliography, tracks references and calculates paper and author ratings.

In addition to explicit references, the system also tracks implicit references. These are similar results and mentions of authors in text without reference. Comparison of automatically extracted results of papers is based on methods of search that are described in Sect. 4.1.

The system also can automatically organize authors in research groups on the basis of statistics of their cooperative publication activity [3]. Authors who has several mutual publications are considered to be working together. The system automatically can distinguish research group leaders, kernel, secondary and invited authors. Leaders of research group often get most portion of citations of the rest of the research group. Kernel authors differ from secondary authors by activity. Invited authors should have some mutual publications with kernel authors of the research group, but must be included in kernel of other research group. The system calculates statistics and ratings of research groups as well.

4.7 Research Topic Detection

The task of research topic detection consists of grouping together papers from one branch of science with similar topics and giving to the built groups some kind of description that reflects their inner essence. Thus, research topic detection is considered as a clusterization task.

Documents for clusterization are represented as sets of keywords and key phrases analogically to the way they are represented in the method for similar document search. The difference lies in the way weights of keywords and key phrases are calculated. We proposed the *TFIDF*-like value, where *TF* was replaced by *ITF* (described in similar document search method) and *IDF* was replaced by value that we called thematic importance characteristic [21, 27]. It is defined as follows.

$$
\Delta \tilde{I}(w, c, \delta) = IDF_N(w, c \backslash \delta) - IDF_N(w, \delta)
$$
$$
\Delta \tilde{I}^+(w, c, \delta) = \Delta \tilde{I}(w, c, \delta) \times H(\Delta \tilde{I}(w, c, \delta))
$$

(2)

w—word; c—general collection set; δ—target group set; H—Heaviside step function; $IDF_N(w, \tau)$—normalized *IDF* of word w in a collection of documents τ.

It is evolution of common *TFIDF* variants, but takes into account how a keyword or key phrase is important for a thematic collection of documents compared to the general collection of all documents.

Clusterization procedure uses top-N keywords and key phrases to describe a document. Our algorithm of clusterization runs incrementally and concurrently on different computational nodes [21]. As a result, it produces groups of documents, which are considered to refer to the same research topic. The produced groups are also described by a set of keywords and key phrases with their weights visualized as a cloud of tags. There is an ability to change granularity of built groups. This feature helps to understand what topics are scattered across publications and analyze them e.g. by counting some statistics about them.

5 Experiment Results

This section will briefly describe some of conducted experiments. The format of the paper does not allow describing every experiment in detail. To get the full description of the experiments you can follow references.

5.1   Semantic Search

The search algorithm of "Exactus Expert" was tested on Russian Information
Retrieval Evaluation Seminar (ROMIP) [11], a competition of Russian search
engines. In many respects, ROMIP seminars are similar to other world information
retrieval events such as TREC, CLEF, NTCIR, etc. As well as TREC, ROMIP has
cycle nature and is overseen by a program committee consisting of representatives
from academia and industry. Several tracks that correspond to different tasks are
conducted. In few years a new task - search in large-scale collections - was conducted on ROMIP. For example, in 2008 there were two collections containing 1.6
and 3.0 million Russian documents. The evaluation procedure changes from year to
year. In general competitors provide search results for a big set of queries of
different types (e.g. about 30,000 queries in 2008), which are compared against the
ground truth. The ground truth consists of the smaller set of queries (e.g. about 500
queries in 2008) randomly chosen from the big set and assessed by some experts.
Several widely known evaluation metrics are used in ROMIP: precision, recall,
11-point TREC precision-recall graph, Bpref etc.

In 2008, our search algorithm showed the highest precision/recall values, in
2009 the algorithm for searching similar documents showed one of the best results.
The experiments show the advantage of using linguistic methods together with
statistical methods for improvement of search quality.

5.2   Search for Similar Documents

Search for similar documents was tested on collection of 30 thousand scientific
publications in the Russian language (20 thousand of scientific papers from leading
Russian journals and 10 thousand published PhD theses abstracts) and Russian
Wikipedia collection of 960 thousand articles. Quality of methods was estimated
using experts-assessors judgment similar to pooling strategy described in [28]. We
used 10 gold standard documents and pool with size of 155 documents. Two
experts were involved. We evaluated quality of similar document search by calculating nDCG and recall values.

The detailed description of the experiment represented in [26]. The best value of
nDCG was 0.98 and recall 0.79. Experiment showed that proposed methods provide good quality of similar document search.

5.3   Term Definition and Result Extraction, Quality
         Assessment

To evaluate quality of term and result extraction we prepared test set of manually
annotated scientific documents. We used about 100 documents to evaluate result

Fig. 3 Results of automatic
quality assignment, terms and
results extraction

> *Science index (from -5 to 5) equals 4.*
> *Interpretation:*
> Text contains   28% of scientific lexis and   1% of quasi -scientific or
> prescientific lexis. References are present . Problem statement is available
> with probability 0.83. Methods described with probability 0.87. Conclusions
> are available with probability 0.55.
>
> *Results:*
> Methodological approach described in the paper   ... was developed for
> accessing large distributed informational systems ... and applied for disaster
> recovery....
>
> *Definitions:*
> Disaster recovery — capability to recover applications and data after
> disaster...
> The process of creation of such model is called as disaster recovery
> modeling...

extraction. The algorithm for retrieving results of research from papers showed value 0.85 for precision on test data with value 0.90 for precision of detecting theoretical or applied results. To evaluate term extraction we used about 30 documents. The precision of definition and term extraction algorithm was 0.84 and the recall was 0.86. The similar experiment is described in [19].

Example of report generated by the system for a paper quality evaluation, terms and results extraction is represented in Fig. 3.

In the report science index of "4" means that the analyzed paper is more likely to be scientific, because it contains big amount of scientific lexis and lesser amount of quasi-scientific lexis. There is also high probability that paper has a problem statement and a methods description. The system is not sure about presence of conclusion.

5.4 Scientific Topics and Research Groups Detection

The experimental study of methods for research topics and research groups detection was carried out on the material of scientific journals, conference proceedings and theses abstracts. This collection consists of more than 100 thousand scientific papers in Russian and English.

The first experiment consisted in the exploration of topics. The experiments showed that some topics are presented mainly in journals, other in theses abstracts or conference proceedings. As a rule, the research topic is presented differently in collections depending on the research area.

Figure 4 shows dynamics of the publications count (in percent) for the topic "Experts systems" for Russian journals and theses abstracts in engineering published in 2007–2012 years. We can see overall reduction of publication activity for this topic. In addition, we can conclude that the topic is mostly presented in theses abstracts.

Fig. 4 Dynamics of the publication activity for the topic "Experts systems"

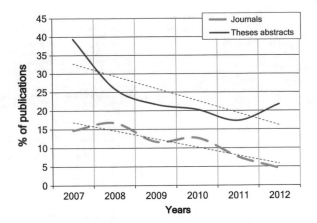

The second experiment consisted in the exploration of research fields. In the engineering area, 32 fields were detected; every field is presented by 35 publications in average.

The third experiment consisted in analyzing research groups. In the field, "Artificial Intelligence" 36 research groups were detected with 6 authors in every research groups in average. Figure 5 shows the publication activity for three selected research groups.

We can see dynamics and the difference in amount of publications for every year. Such figures help evaluate competence of the research groups in the field for searching perspective research groups and making decision about financial support of them.

The system also provides ability to see and interact with graphical representation of relationships between research groups, authors, papers and paper results. Figure 6 shows sample visual representation that can be generated by the system.

A publication title is located in the center of the figure. It is linked with cited publications, the author, which in its turn is linked with a research group. Some

Fig. 5 Dynamics of the publication activity for several research groups

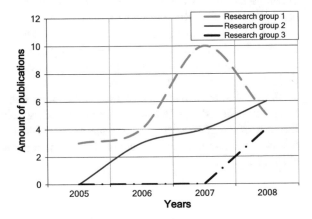

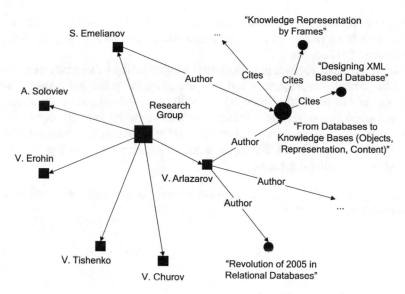

Fig. 6 Example of visual representation of relations between research group authors and publications

other publications are also linked with their authors. It is possible to expand the picture in multiple ways: to display information about publications of other authors, of the research group, expand information about border publications, about their results and so on. This tool helps to surf through authors, publications, research groups and paper results.

6 Conclusion and Future Work

The developed search and analytical engine "Exactus Expert" is a multifunctional tool specially designed to support R&D activities. The system has many features that can help experts and researchers to solve various tasks on exploration of the state of the art in various scientific areas. It can help to detect research topics and examine their structure: distribution of scientific publications of the research topic across different source types (journals, conference papers, PhD theses etc.). The system provides ability to study dynamics of research area development by monitoring publication activity year-by-year. It can detect research groups in a given science field and evaluate their achievements and productivity. The system has ability to find out potential plagiarism and possible duplicates of scientific papers, to track succession (or reveal its absence) in the results of research. It also can evaluate quality of publications, detect scientific, quasi-scientific and unscientific texts.

The search and analytical engine "Exactus Expert" is demanded by experts to support the decision making process on research topics funding by giving aggregated information about different sides of scientific activity. It also can be helpful for editors of science journals and for researchers themselves, especially for Masters and PhD-students.

The experiments showed the possibility to successfully solve many problems of scientific publication processing and analysis using the presented system. Further developments in the field of publication quality analysis are carried on at present. We plan to develop methods for detection of logical defects in scientific papers. We also plan to improve quality of English text analysis and conduct more experiments with English texts. We are looking forward to improve our crawling system and expand our scientific databases.

Acknowledgements The research is supported by Russian Foundation for Basic Research, project No. 14-29-05008-ofi_m.

References

1. SciVal. http://info.scival.com. Cited 15 July 2013
2. R.A. Day, The origins of the scientific paper: the IMRAD format. Am. Med. Writers Assoc. J. **4**(2), 16–18 (1989)
3. D. Deviatkin, A. Shevets, Experimental method of automatic extraction of research topics and research groups, in *Proceedings of Thirteenth National Conference on Artificial Intelligence with international participation CAI-2012*, vol. 2, BGTU, Belgorod, Russia (2012) pp. 90–100. (In Russian)
4. W.B. Frakes, R. Baeza-Yates, *Information retrieval: data structures and algorithms* (Prentice-Hall Inc, Upper Saddle River, NJ, 1992)
5. C. Hui Han, L. Giles, E. Manavoglu, H. Zha, Z. Zhang, E.A. Fox, Automatic document metadata extraction using support vector machines, in *JCDL '03: Proceedings of the 3rd ACM/IEEE-CS Joint Conference on Digital Libraries*. IEEE (2003), pp. 37–48
6. T. Joachims, A probabilistic analysis of the rocchio algorithm with TFIDF for text categorization, *Proceedings of the Fourteenth International Conference on Machine Learning*, ICML '97 (Morgan Kaufmann Publishers Inc., San Francisco, CA, 1997), pp. 143–151
7. L. Leydesdorff, S. Carley, I. Rafols, Global maps of science based on the new Web-of-Science categories. Scientometrics **94**(2), 589–593 (2013)
8. E. Mbaykodzhi, A. Dral, I. Sochenkov, Method for automatic classification of short text messages. Inf. Technol. Comput. Syst. **3**, 93–102 (2012). (In Russian)
9. M.A. Montemurro, Beyond the Zipf-Mandelbrot law in quantitative linguistics. Elsevier (2001), pp. 567–578
10. F. Moya-Anegon, Z. Chinchilla-Rodriguez, B. Vargas-Quesada, E. Corera-Alvarez, F. J. Munoz-Fernandez, A. Gonzalez-Molina, V. Herrero-Solana, Coverage analysis of Scopus: a journal metric approach. Scientometrics **73**(1), 53–78 (2007)
11. M. Nekrestyanov, I. Nekrestyanov, ROMIP 2008 evaluation: rules, methodology and adhoc decisions, in *Proceedings of ROMIP '2008*, NU CZSI, Saint Petersburg, Russia (2008), pp. 5–26. (In Russian)
12. G. Osipov, Formulation of subject domain models: Part 1. Heterogeneous semantic nets. J. Comput. Syst. Sci. Int. **30**(5) (1992)
13. G. Osipov, Methods for extracting semantic types of natural language statements from texts, in *10th IEEE International Symposium on Intelligent Control*. Monterey, California, USA (1995), pp. 292–299.

14. G. Osipov, I. Smirnov, I. Tikhomirov, Relational-situational method for text search and analysis and its applications. Sci. Tech. Inf. Process. **37**(6), 432–437 (2010)
15. G. Osipov, I. Smirnov, I. Tikhomirov, A. Shelmanov, Relational-situational method for intelligent search and analysis of scientific publications, in *Proceedings of the Workshop on Integrating IR technologies for Professional Search, in conjunction with the 35th European Conference on Information Retrieval (ECIR'13)*, vol. 968. (2013), pp. 57–64
16. G. Osipov, I. Smirnov, I. Tikhomirov, O. Zavjalova, Application of linguistic knowledge to search precision improvement, in *Proceedings of 4th International IEEE Conference on Intelligent Systems*, Vol. 2. IEEE (2008), pp. 17-2–17-5
17. G. Osipov, I. Smirnov, I. Tikhomirov, I. Sochenkov, A. Shelmanov, A. Shvets, Information Retrieval for R&D Support // Professional Search in the Modern World. Lect. Notes Comput. Sc. **8830**, 45–69 (2014).
18. L. Padro, E. Stanilovsky, Freeling 3.0: towards wider multilinguality, in *Proceedings of the Language Resources and Evaluation Conference (LREC 2012)*. ELRA, Istanbul, Turkey (2012)
19. A. Shelmanov, Method for automatic extraction of multiword terms from texts of scientific publications, in *Proceedings of Thirteenth National Conference on Artificial Intelligence with international participation CAI-2012*, Vol. 1. BGTU, Belgorod, Russia (2012), pp. 268 274. (In Russian)
20. A. Shvets, A Method of Automatic Detection of Pseudoscientific Publications, in *Proceedings of the 7th IEEE International Conference Intelligent Systems (IS'2014 IEEE). Advances in Intelligent Systems and Computing (AISC)*, Vol. 2. Warsaw (2015), pp. 533–539.
21. A. Shvets, D. Devyatkin, I. Sochenkov, I. Tikhomirov, K. Popov, K. Yarygin, Detection of Current Research Directions Based on Full-Text Clustering, in *Proceedings of Science and Information Conference 2015*, London, (2015), pp. 483–488.
22. I. Smirnov, I. Tikhomirov, Heterogeneous semantic networks for text representation in intelligent search engine EXACTUS, *Proceedings of Workshop SENSE'09—Conceptual Structures for Extracting Natural language SEmantics*, The 17th International Conference on Conceptual Structures (ICCS'09) (CEUR Workshop Proceedings, Moscow, Russia, 2009), pp. 1–9
23. I. Sochenkov, Text comparision method for search and analytical tasks, Artif. Intell. Decis. Mak. **2**, 95–106 (2013). (In Russian)
24. A. Sokirko, A short description of Dialing project (2001). http://www.aot.ru/docs/sokirko/sokirko-candid-eng.html. Cited 15 July 2013
25. L.B. Sollaci, M.G. Pereira, The introduction, methods, results, and discussion (IMRAD) structure: a fifty-year survey. J. Med. Libr. Assoc. **92**(3), 364–371 (2004)
26. R. Suvorov, I. Sochenkov, Detection of similarity of scientific documents using thematic importance characteristic. Artif. Intell. Decis. Mak. **1**, 33–40 (2013). (In Russian)
27. R. Suvorov, I. Sochenkov, I. Tikhomirov, Method for pornography filtering in the web based on automatic classification and natural language processing, in M. Železný et al. (ed.) Lecture Notes in Artificial Intelligence, vol. 8113. (Springer, Berlin 2013), pp. 233–240
28. J. Zobel, How reliable are the results of large-scale information retrieval experiments? in *SIGIR '98: Proceedings of the 21st Annual International ACM SIGIR Conference on Research and Development in Information Retrieval*. ACM, Melbourne, Australia (1998), pp. 307–314
29. G. Zolotova, *Syntactic dictionary: repertory of elementary units of Russian Syntax* (Nauka, Moscow, Russia, 1988). (In Russian)
30. G. Zolotova, N. Onipenko, M. Sidorova, *Communicative grammar of Russian language* (V.V. Vinogradov Russian Language Institute RAS, Moscow, Russia, 2004). (In Russian)

Acoustic and Device Feature Fusion for Load Recognition

Ahmed Zoha, Alexander Gluhak, Michele Nati,
Muhammad Ali Imran and Sutharshan Rajasegarar

Abstract Appliance-specific Load Monitoring (LM) provides a possible solution to the problem of energy conservation which is becoming increasingly challenging, due to growing energy demands within offices and residential spaces. It is essential to perform automatic appliance recognition and monitoring for optimal resource utilization. In this paper, we study the use of non-intrusive LM methods that rely on steady-state appliance signatures for classifying most commonly used office appliances, while demonstrating their limitation in terms of accurately discerning the low-power devices due to overlapping load signatures. We propose a multi-layer decision architecture that makes use of audio features derived from device sounds and fuse it with load signatures acquired from energy meter. For the recognition of device sounds, we perform feature set selection by evaluating the combination of time-domain and FFT-based audio features on the state of the art machine learning algorithms. Further, we demonstrate that our proposed feature set which is a concatenation of device audio feature and load signature significantly improves the device recognition accuracy in comparison to the use of steady-state load signatures only.

Keywords Non-intrusive load monitoring (NILM) · Energy reduction · Energy monitoring · Audio features · Support vector machines (SVM)

A. Zoha (✉) · A. Gluhak · M. Nati · M.A. Imran · S. Rajasegarar
Centre for Communication Systems Research, University of Surrey, GU2 7XH,
Guildford, UK
e-mail: a.zoha@surrey.ac.uk

S. Rajasegarar
e-mail: r.sutharshan@ee.unimelb.edu.au

A. Zoha · A. Gluhak · M. Nati · M.A. Imran · S. Rajasegarar
Department of Electrical and Electronic Engineering, University of Melbourne,
Melbourne, Australia

© Springer International Publishing Switzerland 2016
M. Hadjiski et al. (eds.), *Novel Applications of Intelligent Systems*,
Studies in Computational Intelligence 586, DOI 10.1007/978-3-319-14194-7_15

1 Introduction

The energy consumption in residential spaces and offices is increasing every year [1], which is a growing concern because the energy resources are limited as well as it has negative implications on the environment (e.g. CO_2 emissions). As a result, we see recent initiatives taken by governments across Europe and USA for the large scale deployment of smart meters for improved energy monitoring.

Smart meters however can only measure energy consumption on a house level granularity, providing little information on the breakdown of the energy spent. This has motivated research efforts to develop appliance-level Load Monitoring (LM) techniques such as Non-Intrusive Load Monitoring (NILM) methods to achieve fine-grained energy monitoring. NILM uses aggregated power measurements acquired from a single point (i.e., meter or circuit-level) and further employ feature extraction methods to extract device features. These features are used to model appliance recognition learning algorithms which performs disaggregation of device specific power usage patterns from the acquired measurements. Appliance-level energy feedback is essential for providing meaningful information to the consumers about their energy consumption behavior as well as for the identification of energy hungry devices. But it is still a challenge for NILM solutions to be accurate and cost effective altogether, especially for the case of low-power consumer appliances. This is because low-power consumer appliances exhibit similar steady-state behavior, and cost-effective NILM solutions rely on steady-state energy consumption characteristics to perform load identification.

Motivated by this, in this paper we propose a multi-layer decision architecture that combines appliance acoustic and steady-state energy features to reliably identify low-power office appliances. Our proposed method correlates device energy consumption characteristic and acoustic activity within an office environment to facilitate steady-state NILM method. In particular we make the following contributions with our presented research:

- We report performance comparison of steady-state feature sets used in NILM for discerning low-power office appliances. In our evaluations, we further demonstrate the limitation of using these feature sets for identifying appliances with overlapping load signature.
- We investigate the discriminative ability of time-domain and frequency based audio features for the task of identifying machine and user-generated acoustical events, and further performs optimal model selection for classification.
- We present a multi-layer decision architecture that combines acoustic and device energy features to reliably identify office appliances, providing increased detection performance for low power devices with overlapping load signatures compared to existing load monitoring methods that only rely on steady state signatures derived from energy consumption measurements.

The remainder of the paper is organized as follows. In the next section we review related research work, highlighting the limitation of the current approaches.

In Sect. 3 we provide an introduction to our proposed multi-layer decision framework, whereas we present our experimental evaluations for load identification and acoustic event recognition in Sect. 4. Finally, we provide conclusion and our future work in Sect. 5.

2 Related Work

The existing approaches to appliance load monitoring can be classified into two categories: Intrusive and Non-intrusive load monitoring (NILM) respectively. The intrusive load monitoring makes use of multiple sensors per appliance to track its operational states. Though we can achieve a high accurate measure of device energy usage through this approach but high installation complexity, cost, as well as calibration and data aggregation are outstanding issues that will not favor the use of this technique. In contrast to that the NILM approach as proposed by Hart [2] requires a single sensor to acquire aggregated power measurements. This is followed by a feature extraction process to extract load or device signatures which are used by machine learning algorithms to disaggregate device specific patterns from the aggregated measurements.

The NILM methods are further classified into steady-state and transient methods based on the sampling frequency of the load signatures. The transient methods require high sampling rate for acquiring transient signatures that is shape, size and duration of the transient waveforms, occurs during state transitions. The transient behavior of major appliances is found to be distinct and researchers [3–5] show that transient features are less overlapping in comparison to steady state signatures; providing much higher recognition accuracy for multi-state and low-power consumer appliances. However transient methods suffer from drawbacks such as sensitivity to the wiring architecture of the target environment and expensive hardware due to high sampling rate requirement, which both limit the applicability of this approach [4].

On the other hand the steady state methods characterize appliances based on variations in their steady state signatures such as power (i.e., real power, reactive power), current and voltage signatures. In the market today, there are several commercial solutions (e.g., Plogg, Kill-A-Watt, and Watts Up) available [6], that have inbuilt energy meter which allow extraction of several of steady-state appliance features apart from energy measurement. It has been found out that appliances with ON/OFF operating characteristics (e.g. desk lamp) as well as high-power loads such as water heaters, refrigerators are easy to disaggregate from the aggregated measurements because of their distinct steady-state signatures as reported in [2, 7]. Conversely, most of the low-power consumer appliances as well as multi-state appliances have overlapping device signatures, thus are hard to recognize using steady-state features alone. Norford [8] tried to reduce the ambiguous overlapping of steady-state signatures by using Fast Fourier Transform (FFT) to acquire harmonic content of the input current. This approach however works for a limited number of appliances whereas accurate recognition of low-power consumer appliances still remains a challenge.

The problem can addressed by exploiting side information that is generated as a byproduct of appliance operation such as, [9] developed an automated annotation system *ANNOT* that make use of external information from ambient sensors along with energy features. It is different from our proposed method because the objective of *ANNOT* is to perform automatic annotation of the data, whereas the system requires installation of multiple sensors. Similarly, in [10] a prototype has been developed that make use of audio and steady-state signatures for appliance recognition. However the focus of the work is to identify large consumer appliances that can easily be identified using state of the art methods. In this paper, we have proposed a solution that improves low-cost steady-state NILM method by utilizing side information acquired from detecting acoustic events within an office environment. The fusion of energy and acoustic features facilitates the NILM method to reliably detect low-power consumer appliances with overlapping load signatures, as discussed in the next section.

3   Multi-layer Decision Framework for Smart Sensing

We have already discussed in Sect. 2 that steady-state load signatures due to their low-sampling rate requirement can easily be extracted using cheap energy meters. However, due to the similarity in the steady-state energy consumption pattern of the low-power consumer appliances, NILM methods fails to perform well. Therefore, we proposed a multi-layer decision framework as shown in Fig. 1a, which combines information from energy meter and the audio sensor in order to perform automatic and reliable detection of low-power consumer appliances. We briefly summarize the functionality of each layer

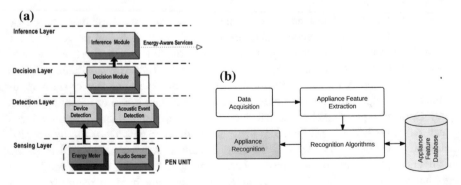

Fig. 1 a Multi-layer decision framework for smart sensing. **b** A general framework for non-intrusive load monitoring

- *Sensing Layer*: The sensing layer consists of an energy meter and an audio sensor installed at a metering point. The aggregated load measurements is acquired from the energy meter whereas the audio sensor provides us with the acoustic information from the environment. The energy consumption pattern of an active appliance in terms of current, voltage, real and reactive power draw, together with acoustic activity in the environment is sensed via the sensing layer. This information is further forwarded to the detection layer for appliance and acoustic event recognition.
- *Detection Layer*: The detection layer consists of device detection and acoustic event detection modules as shown in Fig. 1a.

The functionality of each of these modules is described below.

- *Device Detection Module (DDM)*: The Device Detection Module (DDM) performs automatic appliance recognition using NILM load identification framework as shown in Fig. 1b. The NILM approach is commonly based on supervised learning, that requires extraction of load signatures and correspondingly labelling it with a device class to develop a appliance feature database. The labeled data or the training set is used to train the recognition algorithms so that any test input can be classified by matching it with training examples. The role of DDM is to perform initial detection of appliances based on the steady-state load signatures acquired from the energy meter via the sensing layer. This also involves identification of overlapping appliance classes as demonstrated in Sect. 4.1 and pass on this information to the decision layer in order to perform re-recognition of these confused classes.
- *Acoustic Event Detection Module (AEDM)*: The main functionality of Acoustic Event Detection Module (AEDM) is to perform acoustic event detection for the acoustic surveillance of the target environment. In our experimental evaluations as discussed in Sect. 4.2, we have considered machine and user-generated sounds which most commonly occur within an office environment, as target acoustical events.
- *Decision Layer*: The decision layer reevaluates the recognition results for the confused classes identified by the DDM. In Sect. 4.1 of this paper, we first demonstrate the limitation of using steady-state load signatures in recognizing a set of target office appliances. Furthermore, we addressed this challenge in the decision layer by combining information from AEDM and DDM to improve the appliance recognition.
- *Inference Layer*: The decision layer forward the appliance state information as well as the acoustic cues from the environment to the inference layer. The task of the inference layer is to correlate the acoustic activity of the environment with that of device activity in order to develop energy-aware applications such as user-specific appliance scheduling etc. However, our experimental evaluation in this paper did not demonstrate the functionality of the inference layer which we plan to address in our future work.

4 Experimental Evaluations

For experimental evaluation of our proposed architecture as discussed in Sect. 3, we have considered an office scenario. As for sensing layer, the real-time energy consumption statistics of the nine most common low-power office appliances are collected using Plogg [6] unit. Similarly, 12 target acoustical events which are most likely to occur in an office scenario including user-generated and machine specific sounds are included in the sound database. In order to identify most discriminative feature set for the task of appliance and acoustic event recognition, we have performed feature set and classification model selection experiments in the detection layer. We further analyze the classification performance of each algorithm that enable us to identify which of appliance and sound classes are hard to recognize due to ambiguous overlapping in the feature space. Finally based on the best feature set obtained from DDM and AEDM, we perform feature fusion in the decision layer to improve the low-power device recognition results. Our results and evaluations have been reported in the subsequent sections.

4.1 DDM: Device Recognition Based on Steady-State Load Signatures

4.1.1 Feature Extraction

A load or a device signature is a unique energy consumption pattern of the device that characterizes its operation and distinguishes it from other loads also referred to as device features. The aim of NILM is to perform automatic recognition of the devices and their operational states based on their load signatures. However it requires an appliance feature database to be developed as shown in Fig. 1b, therefore we have collected six steady-state load signatures: real power (P), reactive power (Q), frequency (F), voltage (V_{RMS}), current (I_{RMS}) and phase angle (φ), by measuring the energy consumption target appliances which are listed in Table 1.

Table 1 Target office appliances used for the experimental evaluation

Class no.	Devices	No. of products	Operational states
1	Fluorescent lamp	3	1
2	Incandescent lamp	2	1
3	Laptops	3	3
4	MAC	1	3
5	LCD screen	3	2
6	Fan	2	2
7	Mobile charger	2	1
8	Desktop computer	2	3
9	printer	1	2

Table 2 Device power features used for load recognition

Feature sets	Label	Features
Feature set 1	DF_1	P, Q
Feature set 2	DF_2	I, V, Frequency, Phase angle
Feature set 3	DF_3	I, V, P, Q, Phase angle

We have tried to collect diverse samples of data within each class in order to achieve better generalization. Therefore feature samples from four different laptops, two desktop computers, three mobile chargers, three LCD screens, one printer, one fan and three different lamps have been included in the database. It also allow us to identify intra-class variation of feature values between similar appliance classes. For example, we have found out that the real power consumption of MAC-book is 13 W whereas for other laptops it is around 44 W. Similarly within the lamp class, fluorescent lamps and incandescent lamp show different consumption characteristics. Hence, they are divided into separate device classes. At the end, we have selected nine different appliance classes as listed in Table 1; for which six different load signatures have been collected for each device class. We combine these load signatures into three different feature sets as listed in Table 2.

The selection of features is an important step because it is often a case that the performance of classifier is influenced by the presence of redundant features. In DF_1 we have included two most commonly used device features, real power (P) and reactive (Q), whereas in DF_2 we completely neglected the P and Q features and instead used combination of V_{RMS}, I_{RMS}, Frequency and Phase angle. Finally for DF_3 we have combined DF_1 and DF_2 neglecting the Frequency feature because we have found out from the experimental results that almost all of the devices listed in Table 1 have similar high order harmonics in the range of 49.9–50.1 Hz, thus it has no impact on the classification performance.

4.1.2 Performance Evaluation of Device Feature Sets

The ability of each of these device feature sets (as listed in Table 2) to discriminate between different device classes is evaluated by testing them with the state of the art classification algorithms. k-Nearest Neighbors (k-NN) because of its simplicity and ease of implementation is a popular choice for classification tasks [11], whereas Support Vector Machines (SVM) due to strong mathematical foundation have proven to be effective for text and audio classification [12]. Therefore, we decided to test the performance of each of these classification algorithms against our target dataset. We have used MATLAB simulation environment for offline training and testing of the algorithms. In our implementation of k-NN we use the Euclidean distance metric to partition the multi-class data. On the other hand, SVM is originally a binary class classification algorithm. Therefore, in order to extend it's applicability for multi-class classification problem we have used state of the art one-against one method. We have tested SVM with two kernel functions: polynomial with an

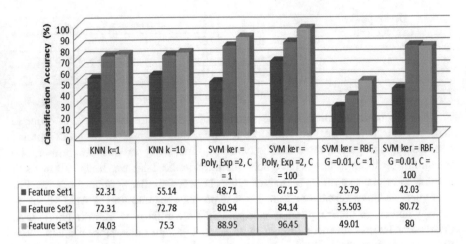

Fig. 2 Appliance recognition accuracy of *k*-NN and SVM models with different configurations using three different device feature sets

exponential value of 2, and Radial Basis Function (RBF) whereas the gamma value (G) of the RBF kernel is found out to be 0.01 using 10-fold cross validation. The overall recognition accuracy of the classifiers with different configurations is shown in Fig. 2.

As for feature set 1, SVM with RBF kernel showed worst performance whereas *k*-NN with *k* value set to 10 show highest classification performance of 55 %. Nevertheless, it has been found out that almost 45 % of instances are misclassified by *k*-NN classifier and only six clusters are formed instead of nine. The combination of *P* and *Q* features usually works well for identifying high power devices because they are well separated in the signature space as reported in [2]. However, low power devices such as lamps, PC, and desk etc. heavily overlap in *P*-*Q* plane, as evident from the results reported in Fig. 2.

Therefore, in feature set 2 we decided to combine I_{RMS}, V_{RMS}, Frequency and Phase angle features. It has already been reported earlier in [8] that I_{RMS} and V_{RMS} show much higher performance in comparison to I_{AVG} and I_{PEAK}. In comparison to feature set 1, SVM and *k*-NN classifiers show significant performance improvement when trained with feature set 2. As for *k*-NN, despite the improvement in recognition accuracy the classes such as fluorescent lamp, laptop, and incandescent lamp are highly overlapped because of similar current and voltage characteristics and hardly 40 % of the instances from these classes are correctly classified. We found no significant improvement in performance of *k*-NN classifier when the value of *k* has been increased from 1 to 10. On the other hand SVM classifier with polynomial kernel shows superior performance over *k*-NN. In contrast to it, SVM with RBF kernel show poor performance; unless the cost function *C* is increased to 100.

The increase in the value of C minimizes the training error forcing the decision boundaries to strictly follow the training data which decreases the generalization capability of the classifier. The consequence would be that the classifier would perform poorly if the test data is even slightly different from the training data which is always the case in a real-world scenario. Therefore, to maintain a good generalization capability of our model we have selected SVM with polynomial kernel function having a cost value of 1. The selected SVM model in combination with feature set 2, despite achieving overall recognition accuracy above 80 %, only less than 60 % of MAC instances are correctly classified because they are confused with incandescent lamp category. Similarly 50 % test instances from LCD screen class are misclassified as laptops. As mentioned earlier, we found out that target appliances selected for this experiment has similar harmonics and hence frequency feature do not aid the classifiers. Therefore, we neglect the frequency feature, and concatenate P, Q, I_{RMS}, V_{RMS}, and phase angle to form feature set 3.

The combination of feature set 3 with SVM classifier (kernel = poly, cost − 1) results in a overall highest recognition accuracy of 89 % as shown in Fig. 2. There is almost an 8 % increase in accuracy in comparison to feature set 2 for the SVM model, whereas we found no significant improvement in results for k-NN classifier. The Confusion Matrix (CM) of our best classification model is shown in Table 3. Each of the device class is represented by a class number as shown in Table 1. Out of 9 device classes, only 5 of them achieves a recognition accuracy above 85 %, whereas MAC, lamp incandescent, Laptop, and LCD Screen show high misclassification rate. The test instances of MAC device class are confused with incandescent lamp device category whereas LCD screen class is confused with laptop and desktop computer as shown in Table 3. In order to minimize the confusion amongst confused classes, we propose a new feature set that increases the separability between the appliance classes in the feature space as discussed in Sect. 4.3.

Table 3 Confusion matrix OF SVM classifier (KER = POLY, COST = 2, C = 1) using FEATURE SET 3

CM	1	2	3	4	5	6	7	8	9
1	**85.3**	2.4	7.3	0	0	5	0	0	0
2	0	**65.7**	0	0	0	0	10	24.3	0
3	14.8	0	**80.2**	0	0	5	0	0	0
4	0	0.8	0	**99.2**	0	0	0	0	0
5	0	0	0	0	**100**	0	0	0	0
6	0	0	4.4	0	0	**95.6**	0	0	0
7	3.32	0	3.31	0	0	0	**93.37**	0	0
8	0	20.8	0	0	0	0	0	**79.2**	0
9	0	0	0	0	0	0	0	0	**100**

4.2 AEDM: Acoustic Event Detection in an Office Environment

As discussed earlier, the role of AEDM is to perform acoustic monitoring of the environment. To demonstrate this, we have included user-specific acoustic events together with machines acoustics in our target sound database. The objective is to test the discriminative ability of the audio feature sets and the classification models for recognizing machine sounds in the presence of user-generated sounds.

4.2.1 Audio Feature Extraction

We have developed an offline classification system, where MATLAB is chosen as a simulation environment for audio feature extraction and classification. We have decided to include 8 user generated and 4 machine generated acoustical events as target sound classes for classification, listed in Table 4.

Due to lack of sound data particularly for machine acoustics, we have obtain sound samples from diverse sources. In addition to acquiring machine sounds using a microphone (indicated by R) as shown in Table 4, we have also obtained samples from the internet (I) and ShATR Multiple Simultaneous Speaker Corpus [13]. Only clapping sound samples belong to Real World Computing Partnership (RWCP) sound scene database [13]. The right most column of Table 4 indicates the number of samples included in the sound database. Each sample has duration of almost 1 min. In the pre-processing step we down sampled sound data to 8 kHz, normalized it in range of [−1 1] and frame-based segmentation is performed (frame length = 128, overlapping 50 % Hamming window). We have removed the silence portion using a threshold based mechanism before extracting the audio features within each frame.

Table 4 Target acoustical events conisdered for the acoustic experiments

No	User-specific AE	Source	Samples
1	Chair moving	I	12
2	Clapping	RWCP + I	100 + 7
3	Cough	I	47
4	Door slam	I	80
5	Laughter	I	26
6	Music	I	38
7	Sneeze	I	40
8	Speech	ShaTR	52
	Machine sounds	Source	Samples
9	Keyboard typing	I	45
10	Printer active mode	I + R	20
11	Mouse click	I + R	13
12	Scanner active mode	I + R	20

Table 5 Audio features

Feature set	Label	Contents	Size
Feature set 1	AF_1	ZCR + STE + F+SLE	8
Feature set 2	AF_2	E + MFCC	13
Fetaure set 3	AF_3	FFBE	13

The time-domain features include Zero Crossing Rate (ZCR), Short Time Energy (STE), Fundamental Frequency (F), and Sub-band log energies (SLE). The details of each of them can be found at [13]. As for frequency based audio features we have computed 12 mel-frequency cepstral coefficients (MFCC) for each frame. The application of a second order filter $H(z) = z - z^{-1}$ at the output of mel-scaled filter bands provide us with Frequency Filter Band Energies (FFBE) which is found out to be more discriminative than MFCC for the task of speech recognition [14]. We have combined time-domain and FFT based features into three feature sets, whereas content and size of each feature vector is shown in Table 5.

4.2.2 Performance Evaluation of Audio Feature Sets

To perform audio feature selection, we trained SVM and Gaussian Mixture Models (GMM) for classifying target acoustical events. The optimal configuration of SVM after experimentation is found to be RBF kernel function with a G value of 0.001. GMM on the other hand tries to estimate the underlying probability density functions of the observations assuming that they can be modeled with mixture of Gaussians. We have tried fixed and variable number of Gaussians per class and the best results were achieved using variable mixture components for each sound class as the amount of data in each class is different. The detail description of GMM algorithm can be found at [15]. Figure 3, clearly shows that, for all three audio feature sets SVM classification model outperforms GMM. The highest overall recognition accuracy achieved is 89 % using frequency based feature set AF_3 in

(a)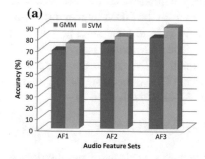

(b)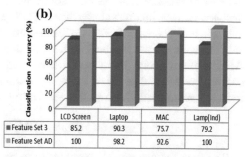

Fig. 3 a A Comparison of SVM and GMM models using different audio feature sets **b** Comparison of F_{AD} and feature Set 3 for classification of confused classes

combination with SVM classifier. However, accuracy is not a true indicator of classifier performance due to data unbalance issue as discussed earlier.

Therefore, in order to analyze the results in detail, we have computed CM and found out that the cough, laughter and sneeze are the most confused classes for the category of user-specific acoustical events. The AF_3 however performs well in comparison to AF_1 and AF_2 in recognizing these confused classes. A comparison of GMM and SVM based classification models for the task of detecting user generated sounds has also been reported in [13], however our experimental evaluation also considers the presence of machine sounds within the environment. Albeit, SVM still outpeforms GMM in our scenario as well. Although, AF_3 achieve an overall high recognition accuracy for all the target sound classes, however from the confusion matrix we found out that printer and scanner sounds are more accurately recognized using feature set AF_2. Conversely, AF_3 performs much better than AF_2 and AF_1 for recognizing keyboard typing and mouse click sound. From this, we can easily conclude that not every feature sets works best for each sound class. The aim of recognizing these machine acoustics is to facilitate the device detection method as discussed below.

4.3 Decision Layer: Feature Fusion of Device and Acoustic Features

In the decision layer, we further improve the classification accuracy of those low-power consumer appliances which are hard to reliably classify due to overlapping of steady-state load signatures as demonstrated in Sect. 4.1. The experimental results obtained in the DDM showed that MAC, Laptop, LCD Screen and Incandescent lamp are the most confused classes. Two of the devices that include MAC and Laptop have a common acoustic signature that is the sound of keyboard typing. This allow us to make an assumption that whenever the user is using a laptop or a MAC device he will generate a keyboard typing sound. Our AEDM can detect and classify this machine sound as already demonstrated in Sect. 4.2. The decision module can reevaluate the result obtained from the DDM by making use of acoustic and energy information. Therefore, we propose to generate a new feature set that concatenate the best audio and device feature acquired from DDM and AEDM respectively. From our previous experiment, SVM (ker = poly, exp = 2, cost = 1) was found out to be the best classifier in the DDM, therefore we decided to train the same model using our new feature set F_{AD} that is defined as

$$F_{AD} = DF_3(from\ DDM) + AF_3(from\ AEDM) \tag{1}$$

The new feature set F_{AD} combines the best feature set selected from the device recognition and the acoustic event recognition experiments respectively. The fusion of acoustic feature with device feature easily separates the MAC and Laptop from incandescent lamp and LCD screen respectively in the feature space. We compared the performance of F_{AD} against DF_3 which is found out to be the best steady-state feature set for device recognition. It is quite evident from the results shown in Fig. 3b that our proposed feature set outperforms feature set 3; increasing the recognition accuracy of the confused classes almost by 16 %.

5 Conclusion and Future Work

This paper has discussed the initial investigation of a multi-layer decision framework for smart energy sensing. The objective is to improve the device recognition accuracy of low-power consumer appliances by combining steady-state load features with audio features derived from the device usage. We investigated the use of time-domain and FFT based audio feature sets for recognizing acoustic activity within an office environment. We found out that FFBE are more discriminative than MFCC based audio features for most of the target sound classes used in the experiment, however few of the machine sounds are best recognized by MFCC based audio feature. SVM was found out to be the best classification model for both audio and device recognition tasks. In future, we will remove our assumptions made in the decision layer by collecting data from real-time experiments. We also plan to implement reasoning strategies in the inference layer in order to perform device usage context recognition.

Acknowledgments We acknowledge the support from the REDUCE project grant (no: EP/I000232/1) under the Digital Economy Programme run by Research Councils UK—a cross council initiative led by EPSRC and contributed to by AHRC, ESRC, and MRC.

References

1. J. Uteley, L. Shorrock, *Domestic Energy Fact File 2008. Technical Representative* (Building Research Establishment, Garston, UK, 2008)
2. G.W. Hart, Nonintrusive appliance load monitoring. IEEE Proc. **80**(12), 1870–1891 (1992)
3. A.I. Cole, A. Albicki, Data extraction for effective non-intrusive identification of residential power loads, in *In Proceedings of Instrumentation and Measurement Technology Conference (IMTC'10)*, vol. 2 (St. Paul, MN, USA, 1998), pp. 812–815
4. Y. Du, L. Du, B. Lu, R. Harley, T. Habetler, A review of identification and monitoring methods for electric loads in commercial and residential buildings, in *In Proceedings of IEEE Energy Conversion Congress and Exposition (ECCE)* (Atlanta,USA, 2010), pp. 4527–4533
5. S.R. Shaw, S.B. Leeb, L.K. Norford, R.W. Cox, Nonintrusive load monitoring and diagnostics in power systems. IEEE Trans. Instrum. Meas. **57**(7), 1445–1454 (2008)
6. M. Hazas, A. Friday, J. Scott, Look back before leaping forward: four decades of domestic energy inquiry. IEEE Pervas. Comput. **10**(1), 13–19 (2011)
7. L.K. Norford, S.B. Leeb, Non-intrusive electrical load monitoring in commercial buildings based on steady-state and transient load-detection algorithms. Energ. Build. **24**(1), 51–64 (1996)

8. T. Kato, H.S. Cho, D. Lee, Appliance recognition from electric current signals for information-energy integrated network in home environments, in *In Proceedings of 7th International Conference on Smart Homes and Health Telematics (ICOST2009)*, vol. 5597. (Springer, Tours, France, 2009), pp. 150–157

9. A. Schoofs, A. Guerrieri, D. Delaney, G. O'Hare, A. Ruzzelli, ANNOT: automated electricity data annotation using wireless sensor networks, in *In Proceedings of 7th Annual IEEE Communications Society Conference on Sensor Mesh and Ad Hoc Communications and Networks (SECON).* (Massachusetts, USA, 2010), pp. 1–9

10. Z.C. Taysi, M.A. Guvensan, Tinyears: spying on house appliances with audio sensor nodes, in *The Proceedings of 2nd ACM Workshop on Embedded Sensing Systems for Energy Efficiency in Building* (2010), pp. 31–36

11. M. Cowling, *Non-Speech Environmental Sound Classification System for Autonomous Surveillance.* Ph.D. thesis (Faculty of Engineering and Information Technology, Griffith University 2004)

12. C.J.C. Burges, A tutorial on support vector machines for pattern recognition. Data Min. Knowl. Disc. **2**(2), 43 (1998)

13. A. Temko, C. Nadeu, Classification of acoustic events using svm-based clustering schemes. Pattern Recogn. **39**, 682–694 (2006)

14. C. Nadeu, J. Hernando, M. Gorricho, On the decorrelation of filter-bank energies in speech recognition, in *Proceedings of Eurospeech* (1995), pp. 1381–1384

15. J.A. Bilmes, A gentle tutorial of the EM algorithm and its application to parameter estimation for gaussian mixture and hidden markov models. Int. Comput. Sci. Inst. **4**(510), 126 (1998)

Printed in the United States
By Bookmasters